Roger Lee (Ed.)

Software Engineering, Artificial Intelligence, Networking and Parallel/Distributed Computing 2010

Studies in Computational Intelligence, Volume 295

Editor-in-Chief

Prof. Janusz Kacprzyk
Systems Research Institute
Polish Academy of Sciences
ul. Newelska 6
01-447 Warsaw
Poland
E-mail: kacprzyk@ibspan.waw.pl

Roger Lee (Ed.)

Software Engineering, Artificial Intelligence, Networking and Parallel/Distributed Computing 2010

Guest Editors

Jixin Ma
Liz Bacon
Wencai Du
Miltos Petridis

 Springer

Prof. Roger Lee
Software Engineering & Information
Technology Institute
Computer Science Department
Central Michigan University
Mt. Pleasant, MI 48859, U.S.A.
E-mail: lee1ry@cmich.edu

ISBN 978-3-642-42245-4   ISBN 978-3-642-13265-0 (eBook)

DOI 10.1007/978-3-642-13265-0

Studies in Computational Intelligence   ISSN 1860-949X

Typeset & Cover Design: Scientific Publishing Services Pvt. Ltd., Chennai, India.

Printed on acid-free paper

9 8 7 6 5 4 3 2 1

springer.com

Preface

The purpose of the 11th Conference on Software Engineering, Artificial Intelligence, Networking, and Parallel/Distributed Computing (SNPD 2010) held on June 9 – 11, 2010 in London, United Kingdom was to bring together researchers and scientists, businessmen and entrepreneurs, teachers and students to discuss the numerous fields of computer science, and to share ideas and information in a meaningful way. Our conference officers selected the best 12 papers from those papers accepted for presentation at the conference in order to publish them in this volume. The papers were chosen based on review scores submitted by members of the program committee, and underwent further rounds of rigorous review.

In Chapter 1, Cai Luyuan et al. Present a new method of shape decomposition based on a refined morphological shape decomposition process.

In Chapter 2, Kazunori Iwata et al. propose a method for reducing the margin of error in effort and error prediction models for embedded software development projects using artificial neural networks (ANNs).

In Chapter 3, Viliam Šimko et al. describe a model-driven tool that allows system code to be generated from use-cases in plain English.

In Chapter 4, Abir Smiti and Zied Elouedi propose a Case Base Maintenance (CBM) method that uses machine learning techniques to preserve the maximum competence of a system.

In Chapter 5, Shagufta Henna and Thomas Erlebach provide a simulation based analysis of some widely used broadcasting schemes within mobile ad hoc networks (MANETs) and propose adaptive extensions to an existing broadcasting algorithm.

In Chapter 6, Qinglei Zhang et al. combine two machine learning techniques – Support Vector Machine (SVM) and Ant Colony Optimization (ACO) – to try to capitalize on the benefits of each and minimize the drawbacks.

In Chapter 7, Harry Goldingay and Jort van Mourik propose an evolutionary multi-agent algorithm to solve a mail problem in various environments.

In Chapter 8, Chowdhury Farhan Ahmed et al. propose a new tree structure for data mining called High Utility Stream tree (HUS-tree) along with a novel algorithm called High Utility Pattern Mining over Stream data (HUPMS) to help solve the problems of existing algorithms.

In Chapter 9, Mohammad Abid Khan et al. present a rule-based algorithm for the resolution of sluices occurring in dialogue.

In Chapter 10, Daniel Demski and Roger Lee explore possible improvements for touchscreen and stylus gesture input functionality through making gestures more context-based, dynamic, and interactive.

In Chapter 11, Alper Ozcan and Sule Gunduz Oguducu design the framework of a recommendation system for mobile service providers, allowing customers to request recommendations from their mobile devices for restaurants, among other possibilities.

In Chapter 12, Jihen Majdoubi et al. propose a contribution for conceptual indexing of medical articles by using the Medical Subject Headings (MeSH) thesaurus and using a language model to index articles with MeSH headings.

It is our sincere hope that this volume provides stimulation and inspiration, and that it will be used as a foundation for works yet to come.

June 2010   Jixin Ma
                                                                            Liz Bacon
                                                                           Wencai Du
                                                                      Miltos Petridis

Contents

List of Contributors

Chowdhury Farhan Ahmed
Kyung Hee University, South Korea
Farhan@khu.ac.kr

Mushtaq Ali
University of Peshawar, Pakistan
mabid@upesh.edu.pk

Yoshiyuki Anan
Omron Software Co., Ltd., Japan
y-anan@mx.omronsoft.co.jp

Tomáš Bureš
Charles University, Czech Republic
bures@dsrg.mff.cuni.cz

Daniel Demski
Central Michigan University, MI,
United States
demsk1da@cmich.edu

Zied Elouedi
Institut Supérieur de Gestion, Tunisia
zied.elouedi@gmx.fr

Thomas Erlebach
University of Leicester, United
Kingdom
t.erlebach@le.ac.uk

Wenying Feng
Trent University, ON, Canada
wfeng@trentu.ca

Faiez Gargouri
Higher Institute of Computer Science
and Multimedia, Tunisia
faiez.gargouri@fsegs.mu.tn

Harry Goldingay
Ashton University, United Kingdom
goldinhj@aston.ac.uk

Shagufta Henna
University of Leicester,
United Kingdom
Sh334@le.ac.uk

Petr Hnětynka
Charles University, Czech Republic
hnetynka@dsrg.mff.cuni.cz

Gongzhu Hu
Central Michigan University,
United States
hu1g@cmich.edu

Naohiro Ishii
Aichi Institute of Technology, Japan
ishii@aitech.ac.jp

Kazunori Iwata
Aichi University, Japan
kazunori@
    vega.aichi-u.ac.jp

Byeong-Soo Jeong
Kyung Hee University, South Korea
jeong@khu.ac.kr

Alamgir Khan
University of Peshawar, Pakistan
mabid@upesh.edu.pk

Mohammad Abid Khan
University of Peshawar, Pakistan
mabid@upesh.edu.pk

Roger Lee
Central Michigan University, MI,
United States
lee1ry@cps.cmich.edu

Cai Luyuan
China University of Geosciences,
China

Jihen Majdoubi
Higher Institute of Computer Science
and Multimedia, Tunisia
Majdoubi_jihene@yahoo.fr

Zhao Meng
Beihang University, China

Jort van Mourik
Ashton University, United Kingdom
vanmourj@aston.ac.uk

Toyoshiro Nakashima
Sugiyama Jogakuen University, Japan
nakasima@sugiyama-u.ac.jp

Sule GunduzOguducu
Istanbul Technical University, Turkey

Alper Ozcan
Istanbul Technical University, Turkey
ozcanalp@itu.edu.tr

Liu Shang
Beihang University, China

Viliam Šimko
Charles University, Czech Republic
simko@dsrg.mff.cuni.cz

Abir Smiti
Institut Supérieur de Gestion, Tunisia
smiti.abir@gmail.com

Syed Khairuzzaman Tanbeer
Kyung Hee University, South Korea
tanbeer@khu.ac.kr

Mohamed Tmar
Higher Institute of Computer Science
and Multimedia, Tunisia
mohamed.tmar@isimsf.mu.tn

Bai Xiao
Beihang University, China

Mao Xiaoyan
Beijing Institute of Control
Engineering, China

Qinglei Zhang
McMaster University, ON, Canada
zhangq33@mcmaster.ca

Shape Decomposition for Graph Representation

Cai Luyuan, Zhao Meng, Liu Shang, Mao Xiaoyan, and Bai Xiao

Abstract. The problem of shape analysis has played an important role in the area of image analysis, computer vision and pattern recognition. In this paper, we present a new method for shape decomposition. The proposed method is based on a refined morphological shape decomposition process. We provide two more analysis for morphological shape decomposition. The first step is scale invariant analysis. We use a scale hierarchy structure to find the invariant parts in all different scale level. The second step is noise deletion. We use graph energy analysis to delete the parts which have minor contribution to the average graph energy. Our methods can solve two problems for morphological decomposition – scale invariant and noise. The refined decomposed shape can then be used to construct a graph structure. We experiment our method on shape analysis.

Keywords: Graph spectra, Image shape analysis, Shape recognition.

1 Introduction

Shape analysis is a fundamental issue in computer vision and pattern recognition. The importance of shape information relies that it usually contains perceptual information, and thus can be used for high level vision and recognition process. There has already many methods for shape analysis. The first part methods can be described as statistical modeling [4] [12][9] [11]. Here a well established route to construct a pattern space for the data–shapes is to

Cai Luyuan
School of Computer Science, China University of Geosciences, Wuhan, China

Zhao Meng, Liu Shang, and Bai Xiao
School of Computer Science and Engineering, Beihang University, Beijing, China

Mao Xiaoyan
Beijing Institute of Control Engineering, Beijing, China

Roger Lee (Ed.): SNPD 2010, SCI 295, pp. 1–10, 2010.
springerlink.com   © Springer-Verlag Berlin Heidelberg 2010

use principal components analysis. This commences by encoding the image data or shape landmarks as a fixed length long vector. The data is then projected into a low-dimensional space by projecting the long vectors onto the leading eigenvectors of the sample covariance matrix. This approach has been proved to be particularly effective, especially for face data and medical images, and has lead to the development of more sophisticated analysis methods capable of dealing with quite complex pattern spaces. However, these methods can't decompose the shapes into parts and can't incorporate high level information from shape. Another problem which may hinder the application of these method is that the encoded shape vectors must be same length which need large human interaction pre-processing.

Another popular way to handle the shape information is to extract the shape skeleton. The idea is to evolve the boundary of an object to a canonical skeletal form using the reaction-diffusion equation. The skeleton represents the singularities in the curve evolution, where inward moving boundaries collide. With the skeleton to hand, then the next step is to devise ways of using it to characterize the shape of the original object boundary. By labeling points on the skeleton using so-called shock-labels, the skeletons can then be abstracted as trees in which the level in the tree is determined by their time of formation[15, 8]. The later the time of formation, and hence their proximity to the center of the shape, the higher the shock in the hierarchy. The shock tree extraction process has been further improved by Torsello and Hancock[16] recently. The new method allows us to distinguish between the main skeletal structure and its ligatures which may be the result of local shape irregularities or noise. Recently, Bai, Latecki and Liu [1] introduced a new skeleton pruning method based on contour partition. The shape contour is obtained by Discrete Curve Evolution [10]. The main idea is to remove all skeleton points whose generating points all lie on the same contour segment. The extracted shape skeleton by using this method can better reflect the origin shape structure.

The previous two shape analysis methods, statistical modeling and shape skeletonization, can be used for shape recognition by combining graph based methods. For example, Luo, Wilson and Hancock [2] show how to construct a linear deformable model for graph structure by performing PCA(Principal Component Analysis) on the vectorised adjacency matrix. The proposed method delivers convincing pattern spaces for graphs extracted from relatively simple images. Bai, Wilson and Hancock[18] has further developed this method by incorporating heat kernel based graph embedding methods. These method can be used for object clustering, motion tracking and image matching. For the shape skeleton methods, the common way is to transform the shape skeleton into a tree representation. The difference between two shape skeletons can be calculated through the edit distance between two shock trees[16].

Graph structure is an important data structure since it can be used to represent the high level vision representation. In our previous work [19], we have introduced an image classifier which can be used to classify image object on different depictions. In that paper, we have introduced an iterative hierarchy image processing which can decompose the object into meaningful parts and hence can be used for graph based representation for recognition.

In this paper, we will introduce a new shape decomposition method. Our method is based on morphological shape decomposition which can decompose the binary shapes through iterative erosion and dilation process. The decomposed parts can then be used to construct a graph structure i.e. each part is a node and the edge relation reflect the relationship between parts, for graph based shape analysis. However, morphological shape decomposition has two shortcomings. First, the decomposition is not scale invariant. When we change the scale level for the same binary shape the decomposition is different. Second, the decomposed parts contains too much noise or unimportant parts. When we use graph based methods for shape analysis these two problems will certainly produce bad influence for our results. Our new method provide two more analysis for morphological decomposition. We first solve the scale invariant problem. We decompose the shape through a hierarchy way. From top to bottom each level representing a different scale size for the same binary shape from small to big. We decompose each level through morphological decomposition and then find the corresponding parts through all levels. We call these parts invariant in all scale levels and use them to represent the binary shapes. The second step is used to delete the noise parts which are normally unimportant and small. We construct the graph structure for the decomposed parts and use graph energy method to analysis the structure. We find the parts(nodes) which has minor or none contribution to the average energy for the whole graph structure. The rest parts are kept as important structure for the shape.

In Section 2, we first review some preliminary shape analysis operations i.e. the tradition morphological shape decomposition. In Section 3, we describe a scale invariant shape parts extraction method. In Section 4, we will describe our graph energy based noise deletion and in Section 5 we provide some experiment results. Finally, in Section 6, we give conclusion and future work.

2   Background on Morphological Shape Decomposition

In this section, we introduce some background on shape morphology operation. Morphological Shape Decomposition (MSD)[14] is used to decompose the shape by the union of all the certain disks contained in the shape. For a common binary shape image, it contains two kinds of elements "0"s and "1"s, where "0" represents backgrounds and "1" represents the shape information. The basic idea of morphology in mathematics can be described as below

$$(M)_u = \{m + u | m \in M\}. \tag{1}$$

There are two basic morphological operations, the dilation of M by S and the erosion of M by S, which are defined as follows:

$$M \oplus S = \bigcup_{s \in S} (M)_S \tag{2}$$

and

$$M \ominus S = \bigcup_{s \in S} (M)_{-S}. \tag{3}$$

There are also two fundamental morphological operation based on dilation and erosion operations, namely the opening of M by $S(M \circ S)$ and closing of M by $S(M \bullet S)$. The definitions are given below:

$$M \circ S = (M \ominus S) \oplus S \tag{4}$$

$$M \bullet S = (M \oplus S) \ominus S \tag{5}$$

A binary shape M can be represented as a union of certain disks contained in M

$$M = \bigcup_{i=0}^{N} L_i \oplus iB \tag{6}$$

where $L_N = X \ominus NB$ and

$$L_i = (M(\bigcup_{j=i+1}^{N})) \ominus iB, \quad 0 \le i < N. \tag{7}$$

N is the largest integer which satisfy

$$M \ominus NB \ne \varnothing,$$

it can be computed by an iterative shape erosion program. B is defined as morphological disks. We call L_i loci and i as corresponding radii. We follow the work by Pitas and Venetsanopoulos [14] to compute the L_i and i. This can give us an initial shape decomposition.

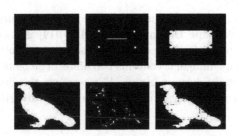

Fig. 1 An example for morphological shape decomposition.

An example is shown in Figure1. Here two shapes(the left column) are given, in which a rectangular shape can be decomposed into five parts. In the upper-middle column of Figure 1 there are one center part and four corners. However, different with the normal shape representation which contains two elements, 0s and 1s, the loci part is represented by the elements of i and the backgrounds are still 0. It is called "Blum Ribbon". With this representation at hand, we can reconstruct the origin shape[14]. The right column in this figure shows the reconstructed shapes by using the "Morphological Ribbon".

3   Scale Invariant Structure Extraction

In the introduction part, we have emphasized the importance of incorporating graph structure representation with shape analysis. It is normal to construct a graph structure from morphological shape decomposition. We can simply treat each part as a node in the graph and the edge relationship is deduced from the adjacency between each pair of parts. If two parts are adjacent or overlap then the weight between the two corresponding nodes are non-zero. In this paper, we dilate the parts with the disk radius size two more than the origin eroded skeleton. For example, if two parts I and J's radius are r_i and r_j with I and J the corresponding loci, we first dilate these two parts by the radius $r_i + 2$ and $r_j + 2$. Then the weight between parts I and J is $and(I \oplus (r_i + 2), J \oplus (r_j + 2))/or(I \oplus (r_i + 2), J \oplus (r_j + 2))$ which reflect both the overlap and adjacent relationship. We can use a five nodes graph to represent the rectangular shape 2 while the center is connected with four corners.

However, the graph structure constructed from this morphological based shape decomposition method is not suitable for graph based shape analysis. It is sensitive to scaling, rotation and noise [7]. An example is shown in Figure 3 here we decompose a set of different size rectangular, we can observe two things 1) It doesn't satisfy scale invariant. As we can see, when the scale is different the decomposition results is different. At the small scale level, the rectangular shape decomposed skeleton include one line and four small triangles. While at large scale level, the skeleton include one line and twelve triangles.

Fig. 2 Graph structure example from morphological shape decomposition.

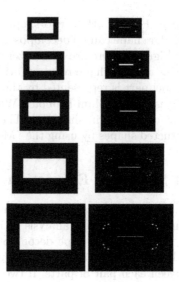

Fig. 3 Example for the same shape morphological decomposition in different scale.

3.1 Hierarchy Morphological Decomposition

We propose a solution which is to decompose the shape in different scale and find the corresponding matching parts to represent the shape. The idea is when a shape is given, we squeeze and enlarge the shape image in a sequence list. We decompose this sequence image shapes. We then find the corresponding parts for this sequence shape decomposition. The stable scale invariant shape decomposition is then found by choose the parts which appear in all different scale levels.

In Figure 3, we still use the example of the rectangular, we first squeeze and enlarge the shape by 15 percent each time. We choose three squeezed and three enlarged shapes – altogether we have five shapes. We then decompose this sequence through morphological decomposition described in the previous section. We then find the correspondence in a hierarchy style. From the correspondence results, we notice that the parts which appear in all levels are the center line and four dots in the corners. The proposed methods can solve the scale invariants problem for shape decomposition. Like SIFT feature [5], we consider the shape decomposition through a hierarchy way.

4 Graph Energy Based Noise Deletion

We continue to use the idea from spectral graph theory [3] to delete the noise in morphological shape decomposition. Our idea is to use graph Laplacian energy which reflect the connectiveness and regularity for the graph to delete the parts(nodes) which has minor or none contribution to the average graph

Laplacian energy per node. The solution is to iteratively delete the parts and finally stop this process when the average graph Laplacian energy per node never rise.

We first review the graph Laplacian energy [6]. The Laplacian matrix is defined as $L = D - A$, in which D is a degree matrix, and A an adjacency matrix. Laplacian graph energy has the following standard definition: for a general graph $G = (V, A)$, with arc weights $w(i, j)$ the Laplacian energy is

$$\mathcal{E}(G) = \sum_{i=1}^{|V|} \left| \lambda_i - 2\frac{m}{|V|} \right| \tag{8}$$

In which: the λ_i are eigenvalues of the Laplacian matrix; m is the sum of the arc weights over the whole graph, or is half the number of edges in an unweighted graph; $|V|$ is the number of nodes in graph. Note that $2m/|V|$ is just the average (weighted) degree of a node. Now, the Laplacian energy of a graph can rise or fall; our tests show that this rise and fall is strongly correlated with the variance in the degree matrix D. This means local minima tend to occur when the graph is regular.

Since we want to use graph Laplaican energy, we need to first construct a graph structure for morphological decomposed parts. The graph structure can be constructed through the method from previous section. We treat each parts from morphology decomposition as a node in the graph G, the edge relationship is found through the adjacency and overlap relationship between each pair of parts.

The process of noise deletion is listed below: 1) We compute the initial average graph energy for the initial state decomposition $\mathcal{E}(G)/|N|$. 2) For each iteration, we go through all the nodes in the graph G. For each node we judge whether we should delete this node. We just compare the previous average graph energy $\mathcal{E}(G)/|N|$ with the average graph energy with this node deleted $\mathcal{E}(G_{di})/|N - i|$, where G_{di} is the graph with ith nodes deleted. If the the average graph energy $\mathcal{E}(G_{di})/|N - i|$ is larger than the previous average energy then we should delete this node and update the graph structure G. 3) Repeat step two, until $\mathcal{E}(G_d i)/|N - i|$ never rise. 4) Output the final decomposition.

The previous process can detect the nodes which has weak link with rest nodes in the graph. It will prune the graph structure until it near or reach regular while keep strong connectiveness within the rest nodes.

5 Experiment

In this section, we provide some experiment results for our methods. Our process can be simply described as below:

- For a given binary shape, we first squeeze and enlarge it to construct the hierarchy scale levels from small to big.

Fig. 4 Sample views of the silhouette objects

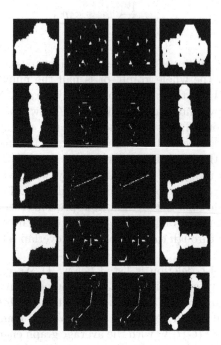

Fig. 5 Example for our methods

- Perform morphological shape decomposition for each level, in this paper we use Pitas and Venetsanopoulos [14] method. Find the corresponding matching parts through all levels. These parts input for the next step.
- Use the output from last step to construct the graph structure. Use average graph energy method to delete the noise nodes(parts) in the graph. Repeat this step until the average graph energy never rise. Output the final graph structure.

We experiment on shock graph database which composed of 150 silhouettes of 10 kinds of objects [16]. An example of database is shown in Figure 4.

In Figure 5, we give some results for our methods, here in the left column is the origin shape, the middle column is the pruned skeleton parts from morphological shape decomposition and the right column is the re-constructed

Table 1 Variation for the number of parts with different shape decomposition methods.

Class Name	MSD	Our Method
Car	8.5	4.1
Children	11.4	6.7
Key	9.0	5.0
Bone	8.5	4.7
Hammer	4.5	3.2

shape by using the skeleton centers in the middle column. From this example, we can see that our algorithm can reduce some noise parts from the origin morphological decomposition while keep the important parts. It can be seen that the reconstructed shapes are quite similar to the original shapes and thus keeps the most important information for further analysis.

In table 1 we listed the variation for the number of parts within the same class for tradition morphological shape decomposition method(MSD) and our method. It is clear that the variations for the number of parts for the tradition morphological shape decomposition is higher than our method.

6 Discussions and Conclusions

In this paper, we proposed a new shape decomposition method which extended the morphological methods. It can conquer two problems for the current morphological methods, scale invariant and noise. We have proposed a graph Laplacian energy based hierarchy shape decomposition. We can extract more stable graph structure by using our methods. Our next step is to use these graph structures to do shape analysis. One possible way is to combine the spectral graph invariants [17] for shape recognition. Recently, Trinh and Kimia [13] has proposed a graph generative for shape through the analysis of shock graphs. We can also extend our methods with graph generative model for morphological decomposition.

References

1. Bai, X., Latecki, L.J., Liu, W.Y.: Skeleton pruning by contour partitioning with discrete curve evolution. IEEE Trans. PAMI 29(3), 449–462 (2007)
2. Luo, B., Wilson, R.C., Hancock, E.R.: A spectral approach to learning structural variations in graphs. Pattern Recognition 39, 1188–1198 (2006)
3. Chung, F.R.K.: Spectral graph theory. American Mathematical Society, Reading (1997)
4. Cootes, T.F., Edwards, G.J., Taylor, C.J.: Active appearance models. In: Burkhardt, H., Neumann, B. (eds.) ECCV 1998. LNCS, vol. 1407, p. 484. Springer, Heidelberg (1998)
5. Lowe, D.: Distinctive image features from scale-invariant keypoints. International Journal of Computer Vision 1, 91–110 (2004)

6. Gutman, I., Zhou, B.: Laplacian energy of a graph. Linear Algebra and its Applications 44, 29–37 (2006)
7. Kim, D.H., Yun, I.D., Lee, S.U.: A new shape decomposition scheme for graph-based representation. Pattern Recognition 38(5), 673–689 (2005)
8. Kimia, B.B., Tannenbaum, A.R., Zucker, S.W.: Shapes, shocks, and deformations. Int. J. Computer Vision 15, 189–224 (1995)
9. Klassen, Srivastava, A., Mio, W., Joshi, S.H.: Analysis of planar shapes using geodesic paths on shape spaces. IEEE Transactions on Pattern Analysis and Machine Intelligence 26, 372–383 (2004)
10. Latecki, L.J., Lakamper, R.: Convexity rule for shape decomposition based on discrete contour evolution. Computer Vision and Image Understanding 77, 441–454 (1999)
11. Lee, C.G., Small, C.G.: Multidimensional scaling of simplex shapes. Pattern Recognition 32, 1601–1613 (1999)
12. Murase, H., Nayar, S.K.: Illumination planning for object recognition using parametric eigenspaces. IEEE Transactions on Pattern Analysis and Machine Intelligence 16, 1219–1227 (1994)
13. Trinh, N., Kimia, B.B.: A symmetry-based generative model for shape. In: International Conference on Computer Vision (2007)
14. Pitas, I., Venetsanopoulos, A.N.: Morphological shape decomposition. IEEE Trans. Pattern Anal. Mach. Intell. 12(1), 38–45 (1990)
15. Shokoufandeh, A., Dickinson, S., Siddiqi, K., Zucker, S.: Indexing using a spectral encoding of topological structure. In: International Conference on Computer Vision and Pattern Recognition, pp. 491–497 (1999)
16. Torsello, A., Hancock, E.R.: A skeletal measure of 2d shape similarity. Computer Vision and Image Understanding 95(1), 1–29 (2004)
17. Xiao, B., Hancock, E.R.: Clustering shapes using heat content invariants, pp. 1169–1172 (2005)
18. Xiao, B., Hancock, E.R.: A spectral generative model for graph structure. In: SSPR/SPR, pp. 173–181 (2006)
19. Xiao, B., Song, Y.-Z., Hall, P.M.: Learning object classes from structure. In: British Machine Vision Conference, Warwich, vol. 1407, pp. 207–217 (2007)

Improving Accuracy of an Artificial Neural Network Model to Predict Effort and Errors in Embedded Software Development Projects

Kazunori Iwata, Toyoshiro Nakashima, Yoshiyuki Anan, and Naohiro Ishii

Abstract. In this paper we propose a method for reducing the margin of error in effort and error prediction models for embedded software development projects using artificial neural networks(ANNs). In addition, we perform an evaluation experiment that uses Welch's t-test to compare the accuracy of the proposed ANN method with that of our original ANN model. The results show that the proposed ANN model is more accurate than the original one in predicting the number of errors for new projects, since the means of the errors in the proposed ANN are statistically significantly lower.

1 Introduction

Due to the expansion in our information-based society, an increasing number of information products are being used. In addition the functionality thereof is becoming

Kazunori Iwata
Dept. of Business Administration, Aichi University
370 Shimizu, Kurozasa-cho, Miyosh, Aichi, 470-0296, Japan
e-mail: kazunori@vega.aichi-u.ac.jp

Toyoshiro Nakashima
Dept. of Culture-Information Studies, Sugiyama Jogakuen University
17-3, Moto-machi, Hoshigaoka, Chikusa-ku, Nagoya, Aichi, 464-8662, Japan
e-mail: nakasima@sugiyama-u.ac.jp

Yoshiyuki Anan
Omron Software Co., Ltd.
Shiokoji Horikawa, Shimogyo-ku, Kyoto, 600-8234, Japan
e-mail: y-anan@mx.omronsoft.co.jp

Naohiro Ishii
Dept. of Information Science, Aichi Institute of Technology
1247 Yachigusa, Yakusa-cho, Toyota, Aichi, 470-0392, Japan
e-mail: ishii@aitech.ac.jp

Roger Lee (Ed.): SNPD 2010, SCI 295, pp. 11–21, 2010.
springerlink.com   © Springer-Verlag Berlin Heidelberg 2010

ever more complex[3, 8]. Guaranteeing software quality is particularly important, because it relates to reliability. It is, therefore, increasingly important for embedded software-development corporations to know how to develop software efficiently, whilst guaranteeing delivery time and quality, and keeping development costs low [2, 6, 7, 9, 10, 12, 13, 14]. Hence, companies and divisions involved in the development of such software are focusing on a variety types of improvements, particularly process improvement. Predicting manpower requirements of new projects and guaranteeing quality of software are especially important, because the prediction relates directly to cost, while the quality reflects on the reliability of the corporation. In the field of embedded software, development techniques, management techniques, tools, testing techniques, reuse techniques, real-time operating systems, and so on, have already been studied. However, there is little research on the relationship between the scale of the development, the amount of effort and the number of errors, based on data accumulated from past projects. Previously, we investigated the prediction of total effort and errors using an artificial neural network (ANN) [4, 5]. In earlier papers, we showed that ANN models are superior to regression analysis models for predicting effort and errors in new projects. In some projects, however, the use of an ANN results in a large margin for error. In this paper, we propose a method for reducing this margin of error and compare the accuracy of the proposed method with that of our original ANN model.

2   Data Sets for Creating Models

Using the following data, we create models to predict both planning effort (Eff) and error (Err).

Eff: "The amount of effort" that needs be predicted.
Err: "The number of errors" in a project.
V_{new}:   "Volume of newly added", which denotes the number of steps in the newly generated functions of the target project.
V_{modify}:   "Volume of modification", which denotes the number of steps modifying and adding to existing functions to use the target project.
V_{survey}:   "Volume of original project", which denotes the original number of steps in the modified functions, and the number of steps deleted from the functions.
V_{reuse}:   "Volume of reuse", which denotes the number of steps in functions of which only an external method has been confirmed and which are applied to the target project design without confirming the internal contents.

3   Artificial Neural Network Model

An artificial neural network (ANN) is essentially a simple mathematical model defining a function.

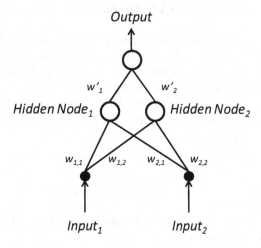

Fig. 1 Basic Artificial Neural Network

$$f : X \rightarrow Y$$

where $X = \{x_i | 0 \le x_i \le 1, i \ge 1\}$ and $Y = \{y_i | 0 \le y_i \le 1, i \ge 1\}$.

ANNs are non-linear statistical data modeling tools that can be used to model complex relationships between inputs and outputs. The basic model is illustrated in Fig. 1, where the output is calculated as follows.

1. Calculate values for hidden nodes. The value of *Hidden Node$_j$* is calculated using the following equation:

$$Hidden\ Node_j = f\left(\sum_i (w_{i,j} \times Input_i)\right) \tag{1}$$

   where $f(x)$ equals $\frac{1}{1+\exp(-x)}$ and the $w_{i,j}$ are weights calculated by the learning algorithm.
2. Calculate *Output* using H_j as follows:

$$Output = f\left(\sum_k (w'_k \times Hidden\ Node_k)\right) \tag{2}$$

   where $f(x)$ equals $\frac{1}{1+\exp(-x)}$ and the w'_k are the weights calculated by the learning algorithm.

We can use an ANN to create efforts and errors prediction models.

3.1 Problems in Original ANN Model

In an ANN, the range of input values or output values is usually less than or equal to 1 and greater than or equal to 0. The values of most selected data, however, are greater than 1. Thus each data range needs to be converted to the range [0, 1] by normalization, a process that leads to a large margin for error in some projects.

In our original ANN model, normalized values are calculated using Eq. (3), where the normalized value for t is expressed as $f_{n_l}(t)$ (where t denotes Eff, Err, V_{new}, V_{modify}, V_{survey}, and V_{reuse}).

$$f_{n_l}(t) = \frac{t - min(T)}{max(T) - min(T)} \tag{3}$$

where T denotes the set of t, and $max(T)$ and $min(T)$ denote the maximum and minimum values of T, respectively.

Since the normalization is flat and smooth, a small change in a normalized value has a greater degree of influence on a small-scale project than on a large scale project.

For example, let $min(T_S)$ equal 10, $max(T_S)$ equal 300, t_{S_1} equal 15, t_{S_2} equal 250, and the predicted values for t_{S_1} and t_{S_2} be $\widehat{t_{S_1}}$ and $\widehat{t_{S_2}}$, respectively. If the prediction model has an error of $+0.01$, then $f_{n_l}^{-1}(0.01) = 2.90$. The predicted values are given as $\widehat{t_{S_1}} = 17.90$ and $\widehat{t_{S_2}} = 252.90$. In both cases the error is the same, but the absolute values of the relative error (ARE) are given by:

$$ARE_{S_1} = \left| \frac{\widehat{t_{S_1}} - t_{S_1}}{t_{S_1}} \right| = \left| \frac{17.90 - 15}{15} \right| = 0.1933$$

$$ARE_{S_2} = \left| \frac{\widehat{t_{S_2}} - t_{S_2}}{t_{S_2}} \right| = \left| \frac{252.90 - 250}{250} \right| = 0.0116$$

The results show that the absolute value of the relative error of the former equation is greater than that of the latter.

The distributions of the amount of effort and the number of errors are shown in Figure 2 and 3, respectively. These distributions confirm that both the amount of effort and number of errors in small-scale projects are major and significant and greater than those in the large scale projects. Thus, in order to improve prediction accuracy, it is important to reconstruct the normalization method.

3.2 Proposed Normalization of Data

In order to solve the problem, we adopt a new normalizing method based on the following equation:

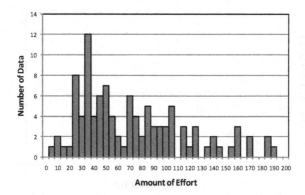

Fig. 2 Distribution of the Amount of Effort

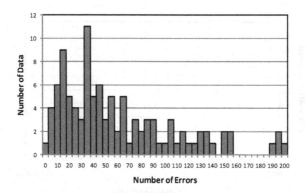

Fig. 3 Distribution of the Number of Errors

$$f_{n_c}(t) = \sqrt{1 - (f_{n_l}(t) - 1)^2} \qquad (4)$$

The comparison between Eqs. (3) and (4) is illustrated in Figures 4 and 5. Equation (4) produces normalized values that increase sharply in the lower range of the original data, and consequently, small changes in the original values are then magnified.

Using the same assumption, the predicted values result in $\widehat{t_{S_1}} = 15.56$ and $\widehat{t_{S_2}} = 271.11$. The absolute values of the relative error are given by Eqs. (5) and (6).

$$ARE_{S_1} = \left| \frac{15.56 - 15}{15} \right| = 0.0373 \qquad (5)$$

$$ARE_{S_2} = \left| \frac{271.11 - 250}{250} \right| = 0.0844 \qquad (6)$$

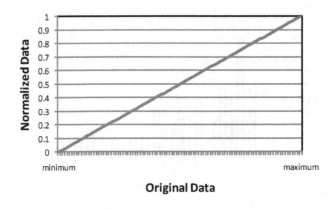

Fig. 4 Results of Normalization using Eq. (3)

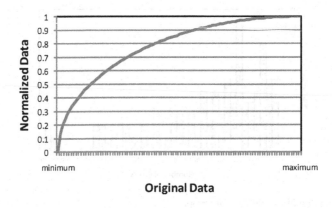

Fig. 5 Results of Normalization using Eq. (4)

The results show that the absolute value of the relative error for small-scale projects is smaller than that produced by our original normalization method and, in contrast, the value for large scale projects is slightly larger than that given by the original normalization method. A more detailed comparison is given in Section 4.

3.2.1 Structure of Model

In a feedforward ANN, the information is moved from input nodes, through the hidden nodes to the output nodes. The number of hidden nodes is important, because if the number is too large, the network will over-train. The number of hidden nodes is generally 2/3 or twice the number of input nodes. In this paper, we use three hidden layers and 36 hidden nodes for each layer in our model as illustrated in Figure 6.

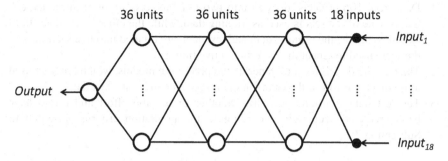

Fig. 6 Structure of Model

4 Evaluation Experiment

4.1 Evaluation Criteria

Eqs. (7) to (10) are used as evaluation criteria for the effort and error prediction models. The smaller the value of each evaluation criterion, the higher is the relative accuracy in Eqs. (7) to (10). The accuracy value is expressed as X, the predicted value as $\widehat{X}$, and the number of data is n.

1. Mean of Absolute Errors (*MAE*).
2. Standard Deviation of Absolute Errors (*SDAE*).
3. Mean of Relative Error (*MRE*).
4. Standard Deviation of Relative Error (*SDRE*).

$$MAE = \frac{1}{n}\sum |\widehat{X} - X| \tag{7}$$

$$SDAE = \sqrt{\frac{1}{n-1}\sum \left(|\widehat{X} - X| - MAE \right)^2} \tag{8}$$

$$MRE = \frac{1}{n}\sum \left| \frac{\widehat{X} - X}{X} \right| \tag{9}$$

$$SDRE = \sqrt{\frac{1}{n-1}\sum \left(\left| \frac{\widehat{X} - X}{X} \right| - MRE \right)^2} \tag{10}$$

4.2 Data Used in Evaluation Experiment

We performed various experiments to evaluate the performance of the original and proposed ANN models in predicting the amount of effort and the number of errors. The experimental conditions are given below:

1. Data from 106 projects, divided into two random sets, are used in the experiments. One of the sets is used as training data, while the other is test data. Both data sets, that is, the training data and test data, are divided into five sections and these are used to repeat the experiment five times.
2. The training data is used to generate the prediction models, which are then used to predict the effort or the errors in the projects included the test data.
3. The test data is used to confirm whether or not the effort and errors have been predicted accurately according to the prediction criteria presented in Subsection 4.1

4.3   Results and Discussion

For each model, averages of the results for the five experiments are shown in Table 1.

Table 1 Experimental Results

	MAE	SDAE	MRE	SDRE
Efforts Prediction				
Original ANN Model	36.02283984	30.90329839	0.673566256	0.723781606
Proposed ANN Model	35.90594900	29.88227232	0.639691515	0.634148798
Errors Prediction				
	MAE	SDAE	MRE	SDRE
Original ANN Model	30.45709728	29.66022495	1.149346138	2.495595739
Proposed ANN Model	28.78197398	27.11388374	0.848232552	1.429305724

4.3.1   Validation Analysis of the Accuracy of the Models

We compare the accuracy of the proposed ANN model with that of the original ANN model using Welch's t-test [15]. The t-test (commonly referred as the Student's t-test) [11] tests the null hypothesis that the means of two normally distributed populations are equal. Welch's t-test is used when the variances of the two samples are assumed to be different and tests the null hypothesis that the means of two not-normally distributed populations are equal if the two sample sizes are equal [1]. The t statistic to test whether the means are different is calculated as follows:

$$t_0 = \frac{|\overline{X} - \overline{Y}|}{\sqrt{\frac{s_x}{n_x} + \frac{s_y}{n_y}}} \qquad (11)$$

where $\overline{X}$ and $\overline{Y}$ are the sample means, s_x and s_y are the sample standard deviations, and n_x and n_y are the sample sizes. For use in significance testing, the distribution of the test statistic is approximated as an ordinary Student's t-distribution with the following degrees of freedom:

$$v = \frac{\left(\frac{s_x}{n_x} + \frac{s_y}{n_y}\right)^2}{\frac{s_x^2}{n_x^2(n_x-1)} + \frac{s_y^2}{n_y^2(n_y-1)}} \tag{12}$$

Thus, once the t-value and degrees of freedom have been determined, a p-value can be found using the table of values from the Student's t-distribution. If the p-value is smaller than or equal to the significance level, then the null hypothesis is rejected.

The results of the t-test for the *MAE* and *MRE* for effort, and the *MAE* and *MRE* for errors are given in Tables 2, 3, 4, and 5, respectively.

The null hypothesis, in these cases, is "there is no difference between the means of the prediction errors for the original and proposed ANN models". The results in Tables 2 and 3 indicate that there is no difference between the respective means of the errors in the predicting effort for original and proposed ANN models, since the p-values are both greater than 0.01 at 0.810880999 and 0.642702832, respectively.

In contrast, the results in Table 4 indicate that the means of the absolute errors are statistically significantly different, since the p-value is 0.000421 and thus less than 0.01. The means of the relative values given in Table 5 also show errors which are statistically significantly different, since the p-value is 0.015592889 and thus greater than 0.01, but less than 0.02. Hence, as a result of fewer errors, the proposed ANN model surpasses the original ANN model in terms of accuracy in predicting the number of errors.

The differences in the results for predicting effort and errors are due to the differences in the distributions shown in Figure 2 and 3. The effort distribution has fewer

Table 2 Results of t-test for *MAE* for Effort

	Original Model	Proposed Model
Mean ($\bar{X}$)	36.02283984	35.90594900
Standard deviation(s)	30.90329839	29.88227232
Sample size (n)	255	255
Degrees of freedom (v)	507.8567107	
T-value (t_0)	0.239414426	
P-value	0.810880999	

Table 3 Results of t-test for *MRE* for Effort

	Original Model	Proposed Model
Mean ($\bar{X}$)	0.673566256	0.639691515
Standard deviation(s)	0.723781606	0.634148798
Sample size (n)	255	255
Degrees of freedom (v)	505.7962891	
T-value (t_0)	0.464202335	
P-value	0.642702832	

Table 4 Results of t-test for *MAE* for Errors

	Original Model	Proposed Model
Mean ($\overline{X}$)	30.45709728	28.78197398
Standard deviation(s)	29.66022495	27.11388374
Sample size (n)	257	257
Degrees of freedom (v)	506.98018	
T-value (t_0)	3.550109157	
P-value	0.000421	

Table 5 Results of t-test for *MRE* Errors

	Original Model	Proposed Model
Mean ($\overline{X}$)	1.149346138	0.848232552
Standard deviation(s)	2.495595739	1.429305724
Sample size (n)	257	257
Degrees of freedom (v)	473.0834801	
T-value (t_0)	2.427091014	
P-value	0.015592889	

small-scale projects than the error distribution and furthermore, the proposed ANN model is more effective for small-scale projects.

5   Conclusion

We proposed a method to reduce the margin of error. In addition, we carried out an evaluation experiment comparing the accuracy of the proposed method with that of our original ANN model using Welch's t-test. The results of the comparison indicate that the proposed ANN model is more accurate than the original model for predicting the number of errors in new projects, the means of the errors in the proposed ANN model are statistically significantly lower. There is, however, no difference between the two models in predicting the amount of effort in new projects. This is due to the effort distribution having fewer small-scale projects than the error distribution and the proposed ANN model being more effective for small-scale projects.

Our future research includes the following:

1. In this study, we used a basic artificial neural network. More complex models need to be considered to improve the accuracy by avoiding over-training.
2. We implemented a model to predict the final amount of effort and number of errors in new projects. It is also important to predict effort and errors mid-way in the development process of a project.

3. We used all the data in implementing the model. However, the data include exceptions and there are harmful to the model. Data needs to be clustered in order to to identify these exceptions.
4. Finally, more data needs to be collected from completed projects.

References

1. Aoki, S.: In testing whether the means of two populations are different (in Japanese), http://aoki2.si.gunma-u.ac.jp/lecture/BF/index.html
2. Boehm, B.: Software engineering. IEEE Trans. Software Eng. C-25(12), 1226–1241 (1976)
3. Hirayama, M.: Current state of embedded software (in japanese). Journal of Information Processing Society of Japan (IPSJ) 45(7), 677–681 (2004)
4. Iwata, K., Anan, Y., Nakashima, T., Ishii, N.: Using an artificial neural network for predicting embedded software development effort. In: Proceedings of 10th ACIS International Conference on Software Engineering, Artificial Intelligence, Nteworking, and Parallel/Distributed Computing – SNPD 2009, pp. 275–280 (2009)
5. Iwata, K., Nakashima, T., Anan, Y., Ishii, N.: Error estimation models integrating previous models and using artificial neural networks for embedded software development projects. In: Proceedings of 20th IEEE International Conference on Tools with Artificial Intelligence, pp. 371–378 (2008)
6. Komiyama, T.: Development of foundation for effective and efficient software process improvement (in japanese). Journal of Information Processing Society of Japan (IPSJ) 44(4), 341–347 (2003)
7. Ubayashi, N.: Modeling techniques for designing embedded software (in japanese). Journal of Information Processing Society of Japan (IPSJ) 45(7), 682–692 (2004)
8. Nakamoto, Y., Takada, H., Tamaru, K.: Current state and trend in embedded systems (in japanese). Journal of Information Processing Society of Japan (IPSJ) 38(10), 871–878 (1997)
9. Nakashima, S.: Introduction to model-checking of embedded software (in japanese). Journal of Information Processing Society of Japan (IPSJ) 45(7), 690–693 (2004)
10. Ogasawara, H., Kojima, S.: Process improvement activities that put importance on stay power (in japanese). Journal of Information Processing Society of Japan (IPSJ) 44(4), 334–340 (2003)
11. Student: The probable error of a mean. Biometrika 6(1), 1–25 (1908)
12. Takagi, Y.: A case study of the success factor in large-scale software system development project (in japanese). Journal of Information Processing Society of Japan (IPSJ) 44(4), 348–356 (2003)
13. Tamaru, K.: Trends in software development platform for embedded systems (in japanese). Journal of Information Processing Society of Japan (IPSJ) 45(7), 699–703 (2004)
14. Watanabe, H.: Product line technology for software development (in japanese). Journal of Information Processing Society of Japan (IPSJ) 45(7), 694–698 (2004)
15. Welch, B.L.: The generalization of student's problem when several different population variances are involved. Biometrika 34(28) (1947)

From Textual Use-Cases to Component-Based Applications

Viliam Šimko, Petr Hnětynka, and Tomáš Bureš

Abstract. A common practice to capture functional requirements of a software system is to utilize use-cases, which are textual descriptions of system usage scenarios written in a natural language. Since the substantial information about the system is captured by the use-cases, it comes as a natural idea to generate from these descriptions the implementation of the system (at least partially). However, the fact that the use-cases are in a natural language makes this task extremely difficult. In this paper, we describe a model-driven tool allowing code of a system to be generated from use-cases in plain English. The tool is based on the model-driven development paradigm, which makes it modular and extensible, so as to allow for use-cases in multiple language styles and generation for different component frameworks.

1 Introduction

When developing a software application, analysts together with end-users negotiate the intended system behaviour. A common practice is to capture system requirements using textual use-cases. Developers then use the use-cases during the coding phase and implement all the business parts of the system accordingly. The business entities identified in the use-cases are usually represented as objects (not necessarily in the sense of the object-oriented approach) with methods/services matching the described actions, variants, extensions etc.

Obviously, there is a significant gap between the original specification, written in natural language, and the final code. When the specified behaviour is manually

Viliam Šimko, Petr Hnětynka, and Tomáš Bureš
Department of Distributed and Dependable Systems
Faculty of Mathematics and Physics, Charles University
Malostranske namesti 25, Prague 1, 118 00, Czech Republic
e-mail: simko,hnetynka,bures@dsrg.mff.cuni.cz

Tomáš Bureš
Institute of Computer Science, Academy of Sciences of the Czech Republic
Pod Vodarenskou vezi 2, Prague 8, 182 07, Czech Republic

Roger Lee (Ed.): SNPD 2010, SCI 295, pp. 23–37, 2010.
springerlink.com   © Springer-Verlag Berlin Heidelberg 2010

coded by the programmer there is a high chance of introducing errors. Apart from that, it is also time- and other resources-consuming task. One of the possible approaches to minimising such human errors is the model-driven development (MDD) [31] where automatic transformation between models is employed. In an ideal case, the last transformation should produce executable code. The code is usually composed of software components [32], since they are well-understood and widely accepted programming technique and offer an explicit architecture, simple reuse, and other advantages.

Even though the model-driven approach is becoming more and more popular, in case of the textual use-cases, there is still a gap that has to be crossed manually, since at the very beginning of the transformation chain there is a specification written in natural language. Furthermore, if textual use-cases come from multiple sources and change in time, they have to be formally verified in order to prevent flaws in the intended specification.

In this paper, we present an automated model-driven approach to generate executable code directly from the textual use-cases written in plain English. The approach is based on our generator (described in [13]), which it further extends and allows for the use-cases to be prepared in different formats and for generating final code in different programming languages and component frameworks.

To achieve the goal the paper is structured as follows. Section 2 provides an overview of the main concepts and tools used in our approach. In Section 3, the transformation process and all used models are described. Section 4 presents related work while Section 5 concludes the paper.

2 Overview

This section gives an overview of concepts and technologies used in our approach. First, we start by introducing main concepts of model-driven development. Then, we briefly elaborate on software components and, finally, we follow by describing textual use-cases and tools for extracting behaviour specification from natural language.

2.1 Model-Driven Development

Using model-driven development, a system is developed as a set of models. Usually, the process starts with modeling the system on a platform independent level, i.e. capturing only the business logic of the system and omitting any implementation details. Then, via series of transformations, the platform independent model is transformed into a platform specific model, i.e. a model capturing also the details specific for the chosen implementation platform.

For model definition, Eclipse Modeling Framework (EMF) [12] is predominantely used nowadays. (It can be viewed as a contemporary de-facto standard.) Models are defined by their meta-models (expressed using EMF), which specify elements that can be used in the corresponding model.

For transformations, the Query-View-Transformation (QVT) language [15] is used. (It is not the only possibility – there are also other transformation languages, e.g. ATL [16], but QVT is the most common.)

2.2 Component-Based Systems

For implementation of applications, MDD usually uses software components, since they offer advantages like explicitly captured application architecture, high level of reuse of already existing components, and others.

Currently, there are a huge number of component frameworks (i.e. system allowing building of application by composition of components) like EJB [23], CCM [14], Koala [25], SCA [8], Fractal [5], SOFA 2 [7], ProCom [6], Palladio [2], and many others. Each of them offers different features, targets different application domains, etc. However, they all view a component as a black-box entity with explicitly specified provided and required services. Components can be bound together only through these services. Some frameworks allow also components to create hierarchies, i.e. to compose component from other components (component frameworks without such a feature allow for a flat composition only).

In general, component-based development has become common for any type of application – not only large enterprise systems but also for software for small embedded and even real-time systems.

2.3 Textual Use-Cases

Use-case specification [19] is an important tool of requirements engineering (RE), which is used for description of intended behavior of a developed system. A use-case is a description of a single task performed in the system. In more detail, it is a description of a process where several entities cooperate together to achieve a goal of the use-case. The entities in the use-case can refer to the whole developed system, parts of the system, or users. A use-case comprises several use-case steps that describe interactions among cooperating entities and that are written in natural language. The use-case is written from the perspective of a single entity called *System Under Discussion* (SuD). The other entities involved in the use-case are (1) a *primary actor*, which is the main subject of SuD communication and (2) *supporting actors* provide additional services to SuD.

The set of use-cases is typically accompanied by a *domain model*, which describes entities appearing in the system being developed. Usually, it is captured as a UML class diagram.

An example of a single use-case is depicted in Figure 1 and a domain model in Figure 2 (both of them are for a marketplace system in detail described in [13]).

To process use-cases, we have previously developed in our research group the Procasor tool (elaborated in [21, 11, 22]). It takes the use-cases in plain English and transforms them into a formal behavior specification called *procases* [26]. The procase is a variant of the *behavior protocol* [27]. In addition to these, UML state

Fig. 1 Example of a textual use-case (also the default format of Procasor)

```
UseCase: Clerk submits an offer on behalf of a Seller

Scope: Marketplace
SuD: Computer System
Primary actor: Clerk
Supporting actor: Trade Commission
Supporting actor: Supervisor
Supporting actor: Seller

Main success scenario specification:
    1. Clerk submits information describing an item
    2. System validates the description
    3. Clerk adjusts/enters price and enters seller's
       contact and billing information
    4. System validates the seller's contact information
    5. System asks the Supervisor to validate the seller
    6. Supervisor permits the seller to operate on the marketplace
    7. System validates the whole offer with the Trade Commission
    8. System lists the offer in published offers
    9. System responds with an uniquely identified
       authorization number

Extensions:
    2a. Validation performed by the system fails
       2a1. Use case aborted
    7a. Trade commission rejects the offer
       7a1. Use case aborted

Sub-variations:
    2b. Price assessment available
       2b1. System provides the seller with a price assessment
```

Fig. 2 Example of a domain model

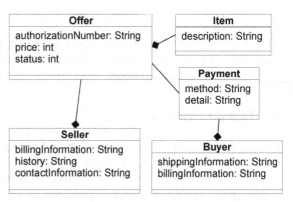

machine diagrams can be also generated. A procase generated from the use-case in Figure 1 is depicted in Figure 3.

In work [13], the Procasor tool has been further extended by a generator that directly generates an executable application from the use-cases. The generator can be seen as an MDD tool. In this view, the use-cases and domain model serve as a platform independent model, which are transformed directly via several transformations into an executable code, i.e. platform specific model. The created application can be used for three purposes: (1) to immediate evaluation of completeness of the use-cases, (2) to obtain first impression by application users, and (3) as a skeleton to be extended into the final application. This intended usage is depicted in Figure 4.

Nevertheless, the generator has several flaws from which the most important is its non-extensibility. It means that the generator allows for a single format of the use-cases. Since it uses the format recommended in the book [19], which is considered as the authoritative guide on how to write use-cases, it is not a major issue. But more importantly, the tool produces JEE (Java Enterprise Edition) applications only.

Fig. 3 Example of a procase

```
?CL.submitItemDescription;
#validateDescription;
#validationPerformedSystemFails;
%ABORT
+
(
        ?CL.submitItemDescription;
        #priceAssessmentAvailable;
        !Sl.providePriceAssessment
        +
        ?CL.submitItemDescription;
        #validateDescription
);
?CL.enterPriceContactBillingInformation;
#validateContactInformation;
!SU.validateSeller;
?SU.permitSeller;
!TC.validateOffer;
(
        #listOffer;
        !Sl.respondUniqAuthNumber
        +
        #tradeCommissionRejectsOffer;
        %ABORT
)
```

In the paper, we address these flaws and propose a real MDD-based system allowing input use-cases in different formats and also allowing generation of application in different component frameworks.

3 Transformation Pipeline

In this section, we present the architecture of our new use-cases-to-code generator. It produces directly executable code form platform independent models, i.e. use-cases. Figure 5 depicts an overview of the generator's architecture.

There are three main parts of the generator. Each of them can be seen as a transformation in MDD sense; an output of one part is an input of the next part. In fact,

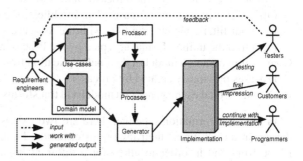

Fig. 4 System usage overview

Fig. 5 Generator architecture

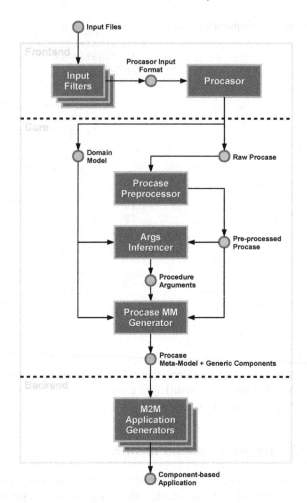

there are additional transformations in each of the parts but they are treated as internal and invisible to the generator users. These parts are:

Front-end: It consists of the Procasor tool however modified in order to accept different formats of the use-cases. This modification can be viewed as the support for input filters, which take files with the use-case specification and transform them into the unified format accepted by the Procasor.

Core: It generates a model of the final application from the domain model and the procases previously created by Procasor.

Back-end: It takes the generated model and transforms it into the executable code.

In this transformation pipeline, there are two extension points where users of the generator can insert their own transformation. First, these are the input filters in the front-end in order to allow for a different input format, and, second, the transformations in the back-end in order to add support of different programming languages and platforms, in which the final application can be produced.

In the following sections, we describe all these parts in more detail.

3.1 Front-End: From Text to Procases

The original Procasor tool accepts use-cases only in the format recommended by [19]. In this format, a single use-case consists of several parts: (1) a header with the use-case name and identification of SuD, PA and supporting actors, (2) the success scenario, which is a sequence of steps performed to achieve the goal of the use-case, (3) extensions and/or sub-variants that are performed instead or in addition to a step from the success scenario. The use-case example already shown in Figure 1 has this format.

Nevertheless, the structure described above is not the only one possible. Therefore, our generator allows for adding support of different use-case formats. These filters convert use-cases into the unified format understood be the Procasor. They are intended to be added by the generator users in order to support their format. Currently, the filters are written as Java classes rather than QVT (or other) transformation since the input are plain text files (i.e. the use-cases) and the Procasor input format (i.e. output format of the filters) is also textual.

Once the Procasor has the use-cases in format it understands, it generates the corresponding procases. At this point, the generated procases can be inspected by users in order to identify potential ambiguities in the use-cases that could "confuse" the Procasor and then to repair the use-cases and regenerate procases.

3.2 Core: From Procases to Components

In this part, the procases and domain model are processed and transformed into elements suitable for the final code generation. The results are *procase model* and *generic component model*; their meta-models are described later in this section.

Internally, it is not a single transformation but rather a sequence of several transformations. However, they are treated as internal and their results are not intended for direct usage. These internal transformations are fully automated and they are as follows.

1. Identification of procedures in the generated procases,
2. Identification of arguments of the procedures,
3. Generating procase models and component models.

The first two steps are in detail described in [13], therefore we just briefly outline them here.

In the first step, procedures are identified in the generated procases. A procedure is a sequence of actions, which starts with a request receive action and continues with other action than request receive. In the final code, these procedures correspond to the methods of generated objects.

Fig. 6 Procase Meta-model

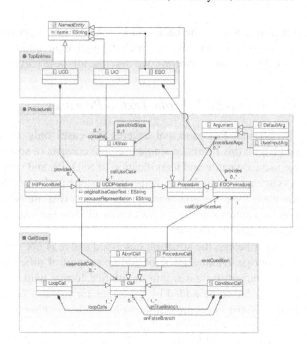

Next, as the second step, arguments of identified procedures and data-flow among them have to be inferred from the domain model. These arguments eventually result into methods arguments in the final code and also into objects' allocations.

Finally, as the third step, procase models and component models are generated from the modified procases and domain model. This step is completely different to the original generator, which would generate JEE code directly.

The procase model represents the modified procases in the MDD style and also covers elements from the domain model. The generic component model contains identified components. Both the models have a schema described by meta-models; the procase meta-model is depicted in Figure 6 and a core of the generic component meta-model in Figure 7.

First, we describe the procase meta-model. It contains following main entities:

- Entity Data Objects,
- Use-Case Objects,
- User-interface steps,
- Procedures (signatures),
- Call-graph (tree) reflecting the use-case flow.

All entities from the domain model (Entity Data Objects) are modeled as instances of the *EDO* class. They contain *EDOProcedures*, i.e. placeholders for the internal logic of actions. Since the domain model and the use-cases do not provide enough information to generate the internal logic of the domain entities (e.g. validation of an item), we generate only skeletons for them. The skeletons contain code which logs their use. This means that the generated application can be launched

Fig. 7 Generic Component
Meta-model

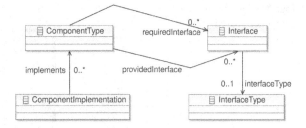

without any manual modification and by itself provides traces, that give valuable feedback when designing the application.

Use-case objects (modeled as the *UCO* class) contain the business logic of the corresponding use-cases, i.e. the sequence of actions in the use-case success scenario and all possible alternatives.

User interface objects (the *UIO* class) represent the interactive part of the final application. Human users can interact via them with the application, set required inputs, and obtain computed results.

Procedures from the procases are modeled as instances of the *UCOProcedure* class, whereas the business logic is modeled as a tree of calls (the *Call* class) with other procedure calls, branches, and loops. More precisely, the following types of calls are supported:

- *ProcedureCall* that represents a call to an *EDOProcedure*.
- *ConditionCall* that represents branching. Depending on the result from the *evalCondition* call to an *EDOProcedure* it either follows the *onTrueBranch*, modeled as a non-empty sequence of calls, or the *onFalseBranch*, modeled as an optional sequence of calls.
- *LoopCall* that represents a loop.
- *AbortCall* that represents forced end of the call.

To preserve the order of calls within a sequence of calls, there is an ordered association in the meta-model.

The text of the original use-case is also captured in the *UCOProcedure* and it can be used during the code generation, e.g., to display it in the user interface with currently executed action highlighted (as it is done in our implementation).

The generic component meta-model is heavily inspired by our SOFA 2 component model [7]; in fact it is a simplified version of the SOFA 2 component model without SOFA 2 specific elements. A component is captured by its component type and component implementation. The type is defined by a set of provided and required interfaces while the implementation implements the type (a single type or several of them) and can be either primitive or hierarchical (i.e. composed of other components).

Component implementations has a reference to the *Procedure* class in the procase meta-model; this reference specifies, which actions belong to the component. Interconnections among components (i.e. required-to-provided interface bindings) are identified based on the actual procedure-calls stored in the procase model.

Fig. 8 Example of a
component-based appli-
cation

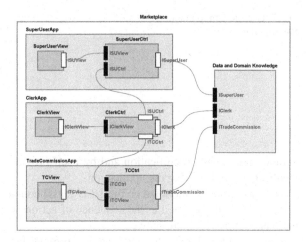

Figure 8 shows an example generated components and their architecture (the
example is the component-based version of the Marketplace example used in [13]).

As depicted, the component structure reflects the actors and the MVC pattern.
For each actor (i.e. the UIO class), there is a composite component with two sub-
components – one corresponding to the view part, another to the controller part. The
controller sub-component represents the logic as defined by the use-cases. It reflects
flows from the use-cases (i.e. UCO classes) in which the particular actor appears as
the primary actor.

The last component of the structure is Data and Domain Knowledge. It contains data-oriented functionality, which cannot be automatically derived from the
use-cases (e.g. validation of an item description). The interfaces of the component correspond to EDO classes (i.e. collections of "black-box" actions used in the
use-cases).

3.3   Back-End: From Model to Application

The back-end part takes the procase model and component model as the input
and generates the final application. It is intended that users of our tool will create back-ends for their required component frameworks. These back-ends can be
implemented as model-to-model transformations (e.g. using QVT) in case of creating final component model and/or as model-to-text transformations (e.g. using the
Acceleo technology [24]) in case of generation of final code or, of course, as plain
Java code.

Currently, we have been working on back-ends for JEE and SOFA 2 (under development); both of them described in the following paragraphs. They can be used
as examples of back-ends generating 1) a flat component model (JEE), and 2) a
hierarchical component model (SOFA 2).

The JEE back-end is an updated version from the original generator (see [13]). It generates the final application as a set of JEE entities organized in the following tiers:

Presentation tier: composed of several JSF (Java Server Faces) pages together with their backing beans.

Use-case objects: representing the actual use-case flow.

Entity data objects: representing the business services of the designed application. The actual code of EDOs has to be provided by developers; the generators implements them via logging (see Section 3.2).

The SOFA 2 back-end has been implemented using QVT transformation and Acceleo templates. The definition of SOFA 2 components is obtained from the generic component model by rather a straightforward mapping (the SOFA 2 frame corresponds to the component type, the SOFA 2 architecture to the component implementation, and interfaces to the interfaces). The code of components is generated from the procase model using Acceleo templates as follows.

- Use-case flow of every *UCOProcedure* is mapped to a sequence of Java commands using the same algorithm as in [13].
- Every *EDOProcedure* is mapped to a method that sends a trace-message the logger.
- *UISteps* representing the user interface are generated as simple forms using the standard Java Swing toolkit in a similar fashion as JSF pages generated in [13].

A deployment plan that defines assignment of instantiated component to runtime containers is also generated. For the sake of simplicity, the default deployment plan uses a single container for all components, nevertheless, it can be easily changed by the user.

4 Related Work

Related work is divided into two sections – first, related to the processing of a natural language and second, related to processing of controlled languages, which is a restricted subset of a natural language in order to simplify its processing.

4.1 Natural Language Processing

In [29], authors elicit user requirements from natural language and create a high-level contextual view on system actors and services – Context Use-Case Model (CUCM). Their method uses the Recursive Object Model (ROM) [33] which basically captures the linguistic structure of the text. Then, a metric-based partitioning algorithm is used to identify actors and services using information from the domain-model. The method generates two artefacts – (1) a graphical diagram showing actors and services and (2) textual use-case descriptions. Similarly to this method, we build a common model capturing the behaviour of use-cases in a semi-automatic way. Rather than rendering graphical diagrams, we focus on code generation. Unlike [29], where every sentence is transformed to a use-case, we divide the text into use-cases and further into use-case steps. Such a granularity is necessary when identifying components and modelling the dependencies among them.

An automatic extraction of object and data models from a broad range of textual requirements is also presented in [20]. This method also requires a manual specification of domain entities prior to the actual language processing. The employed dictionary-based identification approach is, however, not as powerful as the linguistic tools inside Procasor.

In [1], an environment for analysis of natural language requirements is presented. The result of all the transformations is a set of models for the requirements document, and for the requirements writing process. Also the transformation to ER diagrams, UML class diagrams and Abstract State Machines specifying the behaviour of modeled subsystems (agents) is provided. Unlike our generator, however, CIRCE does not generate an executable applications, just code fragments.

4.2 Controlled Languages

In [28], a Requirements-to-Design-to-Code (R2D2C) method and a prototype tool is presented. Authors demonstrate a MDA-compliant approach to development of autonomous ground control system for NASA directly from textual requirements. Their method relies on a special English-like syntax of the input text defined using the ANTLR[1] grammar. User requirements are first transformed to CSP (Hoare's language of Communicating Sequential Processes) and then Java code is generated. Unlike the constrained ANTLR-based grammar, our method benefits from the advanced linguistic tools inside Procasor. Moreover, our generator is not tied to Java and we can generate applications using different component-based systems.

More research on controlled languages and templates have been conducted in the domain of requirements engineering [3, 4, 18, 10, 30] and artificial intelligence [17, 9]. Authors of [30] describe the PROPEL tool that uses language templates to capture requirements. The templates permit a limited number of highly structured phrases that correspond to formal properties used in finite-state automata. In [18], a structured English grammar is employed with special operators tailored to the specification of realtime properties. Similarily, in [10], natural language patterns are used to capture conditional, temporal, and functional requirements statements. Due to the use of controlled languages, these methods require closer cooperation with a human user than is necessary with Procasor. Usually, the goal of these methods is to formally capture functional requirements. Although they cover a broader range for functional requirements than we do, these methods do not produce any executable result.

---

[1] http://www.antlr.org

5 Conclusion

In this paper, we have described next version of a tool that automatically generates an executable component application from textual use-cases written in plain English. Our approach brings important benefits to the software development process, because it allows designs to be tested and code to be automatically regenerated as the result of adjustments in the textual specification. Thus, it naturally ensures correspondence between the specification and the implementation, which often tends to be a challenging task.

Compared to the previous version, the tool strongly relies on MDD and model transformations. This gives it a very strong and extensible architecture, which allows for easy extensions in order to support different use-case formats and component frameworks for the final generated application.

Currently, we have specified all meta-models and we have an implementation able to generate JEE applications (SOFA 2 back-end is under development). Further development needs to be conducted to integrated the tool into Eclipse.

As a future work, we intend to implement more transformation back-ends for other component frameworks (e.g. Fractal component model [5]) and different formats of the use-cases. Another interesting direction of our future work is a support of deriving the domain model also from the textual specification in natural language.

Acknowledgements. This work was partially supported by the Ministry of Education of the Czech Republic (grant MSM0021620838).

References

1. Ambriola, V., Gervasi, V.: On the systematic analysis of natural language requirements with circe. Automated Software Engineering (2006)
2. Becker, S., Koziolek, H., Reussner, R.: The Palladio component model for model-driven performance prediction. Journal of Systems and Software 82, 3–22 (2009)
3. Breaux, T., Anton, A.: Analyzing goal semantics for rights, permissions and obligations. In: RE 2005: Proceedings of the 13th IEEE International Conference on Requirements Engineering, pp. 177–188. IEEE Computer Society, Washington (2005), http://dx.doi.org/10.1109/RE.2005.12
4. Breaux, T., Anton, A.: Deriving semantic models from privacy policies. In: POLICY 2005: Proceedings of the Sixth IEEE International Workshop on Policies for Distributed Systems and Networks, pp. 67–76. IEEE Computer Society, Washington (2005), http://dx.doi.org/10.1109/POLICY.2005.12
5. Bruneton, E., Coupaye, T., Stefani, J.B.: The Fractal Component Model, http://fractal.ow2.org/specification/
6. Bures, T., Carlson, J., Crnkovic, I., Sentilles, S., Vulgarakis, A.: ProCom - the Progress Component Model Reference Manual, version 1.0. Technical Report, Mälardalen University (2008), http://www.mrtc.mdh.se/index.php?choice=publications&id=1508

7. Bures, T., Hnetynka, P., Plasil, F.: SOFA 2.0: Balancing Advanced Features in a Hierarchical Component Model. In: Proceedings of SERA 2006, Seattle, USA, pp. 40–48 (2006)
8. Open SOA Collaboration: SCA Service Component Architecture: Assembly Model Specification (2007), http://www.osoa.org/download/attachments/35/SCA_AssemblyModel_V100.pdf?version=1
9. Cregan, A., Schwitter, R., Meyer, T.: Sydney OWL syntax-towards a controlled natural language syntax for OWL 1.1. In: Proceedings of the OWLED 2007 Workshop on OWL: Experiences and Directions, Innsbruck, Austria, vol. 258 (2007)
10. Denger, C., Berry, D., Kamsties, E.: Higher quality requirements specifications through natural language patterns. In: Proceedings IEEE International Conference on Software - Science, Technology and Engineering, pp. 80–91. IEEE Computer Society, Los Alamitos (2003)
11. Drazan, J., Vladimir, M.: Improved processing of textual use cases: Deriving behavior specifications. In: van Leeuwen, J., Italiano, G.F., van der Hoek, W., Meinel, C., Sack, H., Plášil, F. (eds.) SOFSEM 2007. LNCS, vol. 4362, pp. 856–868. Springer, Heidelberg (2007)
12. Eclipse.org: Eclipse Modelling Framework, http://www.eclipse.org/emf
13. Francu, J., Hnetynka, P.: Automated Code Generation from System Requirements in Natural Language. e-Informatica Software Engineering Journal 3(1), 72–88 (2009)
14. Object Management Group: CORBA Component Model Specification, Version 4.0 (2006)
15. Object Management Group: Meta Object Facility (MOF) 2.0 Query/View/ Transformation Specification (2008)
16. Jouault, F., Allilaire, F., Bezivin, J., Kurtev, I., Valduriez, P.: ATL: a QVT-like transformation language. In: OOPSLA Companion, pp. 719–720 (2006)
17. Kaljurand, K., Fuchs, N.E.: Verbalizing Owl in Attempto Controlled English. In: Proceedings of the OWLED 2007 Workshop on OWL: Experiences and Directions, Innsbruck, Austria, vol. 258 (2007)
18. Konrad, S., Cheng, B.: Real-time specification patterns. In: ICSE 2005: Proceedings of the 27th international conference on Software engineering, pp. 372–381. ACM, New York (2005), http://doi.acm.org/10.1145/1062455.1062526
19. Larman, C.: Applying UML and Patterns: An Introduction to Object-Oriented Analysis and Design and Iterative Development, 3rd edn. Prentice Hall PTR, Upper Saddle River (2004)
20. MacDonell, S.G., Min, K., Connor, A.M.: Autonomous requirements specification processing using natural language processing. In: Proceedings of the ISCA 14th International Conference on Intelligent and Adaptive Systems and Software Engineering (IASSE 2005), pp. 266–270. ISCA, Toronto (2005)
21. Mencl, V.: Deriving behavior specifications from textual use cases. In: Proceedings of Workshop on Intelligent Technologies for Software Engineering (2004)
22. Mencl, V., Francu, J., Ondrusek, J., Fiedler, M., Plsek, A.: Procasor environment: Interactive environment for requirement specification (2005), http://dsrg.mff.cuni.cz/~mencl/procasor-env/
23. Oracle (Sun Microsystems): Java Platform, Enterprise Edition (Java EE): Enterprise JavaBeans Technology, http://java.sun.com/products/ejb/
24. Obeo: Acceleo: Open source plugin for model to text transformation based on templates, http://www.acceleo.org
25. van Ommering, R., van der Linden, F., Kramer, J., Magee, J.: The Koala Component Model for Consumer Electronics Software. Computer 33(3), 78–85 (2000), http://dx.doi.org/10.1109/2.825699

26. Plasil, F., Mencl, V.: Getting 'whole picture' behavior in a use case model. Journal of Integrated Design and Process Science 7(4), 63–79 (2003)
27. Plasil, F., Visnovsky, S.: Behavior protocols for software components. IEEE Transactions on Software Engineering 28(11), 1056–1076 (2002), http://doi. ieeecomputersociety.org/10.1109/TSE.2002.1049404
28. Rash, J.L., Hinchey, M.G., Rouff, C.A., Gracanin, D., Erickson, J.: A requirements-based programming approach to developing a NASA autonomous ground control system. Artif. Intell. Rev. 25(4), 285–297 (2007), http://dx.doi.org/10.1007/ s10462-007-9029-2
29. Seresh, S.M., Ormandjieva, O.: Automated assistance for use cases elicitation from user requirements text. In: Proceedings of the 11th Workshop on Requirements Engineering (WER 2008), 16, Barcelona, Spain, pp. 128–139 (2008)
30. Smith, R., Avrunin, G., Clarke, L., Osterweil, L.: Propel: An approach supporting property elucidation. In: 24th Intl. Conf. on Software Engineering, pp. 11–21. ACM Press, New York (2002)
31. Stahl, T., Voelter, M., Czarnecki, K.: Model-Driven Software Development: Technology, Engineering, Management. John Wiley & Sons, Chichester (2006)
32. Szyperski, C.: Component Software: Beyond Object-Oriented Programming, 2nd edn. Addison-Wesley Professional, Reading (2002) (Hardcover)
33. Zeng, Y.: Recursive object model (ROM)-Modelling of linguistic information in engineering design. Comput. Ind. 59(6), 612–625 (2008), http://dx.doi.org/10. 1016/j.compind.2008.03.002

COID: Maintaining Case Method Based on Clustering, Outliers and Internal Detection

Abir Smiti and Zied Elouedi

Abstract. Case-Based Reasoning (CBR) suffers, like the majority of systems, from a large storage requirement and a slow query execution time, especially when dealing with a large case base. As a result, there has been a significant increase in the research area of Case Base Maintenance (CBM).

This paper proposes a case-base maintenance method based on the machine-learning techniques, it is able to maintain the case bases by reducing its size and preserving maximum competence of the system. The main purpose of our method is to apply clustering analysis to a large case base and efficiently build natural clusters of cases which are smaller in size and can easily use simpler maintenance operations. For each cluster we reduce as much as possible, the size of the cluster.

1 Introduction

Case Based Reasoning [1] [2] [3] is one of the most successful applied machine learning technologies. It is an approach to model the human way in reasoning and thinking. It offers a technique based on reusing past problem solving experiences to find solutions for future problems.

A number of CBR applications have become available in different fields like medicine, law, management, financial, e-commerce, etc. For instance for the e-commerce fields, CBR has been used as an assistant in e-commerce stores and as a reasoning agent for online technical support, as well as an intelligent assistant for sale support or for e-commerce travel agents. It uses cases to describe commodities on sale and identifies the case configuration that meets the customers' requirements [4].

Abir Smiti
LARODEC, Institut Supérieur de Gestion, Tunis Tunsie
e-mail: smiti.abir@gmail.com

Zied Elouedi
LARODEC, Institut Supérieur de Gestion, Tunis Tunsie
e-mail: zied.elouedi@gmx.fr

Roger Lee (Ed.): SNPD 2010, SCI 295, pp. 39–52, 2010.

The CBR cycle describes how to design and implement CBR solutions for successful applications. It starts with researching into previous cases, revising the solution, repairing case and adding it into the case base [5].

The success of a CBR system depends on the quality of case data and the speed of the retrieval process that can be expensive in time especially when the number of cases gets large. To guarantee this quality, maintenance of CBR systems becomes necessarily. According to Leake and Wilson [6], Case Base Maintenance (CBM) is a process of refining a CBR system's case base to improve the system's performance. It implements policies for revising the organization or contents (representation, domain content, accounting information, or implementation) of the case-base in order to facilitate future reasoning for a particular set of performance objectives.

Various case base maintenance policies have been proposed to maintain the case base, most of them are case-deletion policies to control case-base growth [7]. Other works have chosen to build a new structure for the case-base [1]. However, many of them are expensive to run for large CB, and suffer from the decrease of competence especially when it exists some noisy cases, since the competence depends on the type of the cases stored.

We propose in this paper a new method COID - Clustering, Outliers and Internal cases Detection- that could be able to maintain the case bases by reducing its size, and consequently reducing the case retrieval time. We apply clustering analysis to a large case base and efficiently build natural clusters of cases to allow the case base to be maintained at each small case base by selecting and removing cases by preserving maximum competence. In this way, we will obtain a case base size reducing by guiding the CB towards an optimal configuration.

This paper is organized as follows: In Section 2, CBR system and its cycle will be presented. In Section 3, some of strategies for maintenance and quality criteria of the case base will be approached. Section 4 describes in detail our new approach COID for maintaining case base. Finally, Section 5 presents and analyzes experimental results carried out on data sets from the U.C.I. repository [8].

2   The CBR Cycle

The CBR approach has been used in various fields, it is able to find a solution to a problem by employing its luggage of knowledge or experiences which are presented in form of cases. Typically, the case is represented as a pair "problem" and "solution".

To solve the problems, CBR system calls the past cases, it reminds to the similar situations already met. Then, it compares them with the current situation to build a new solution which, in turn, will be added to the case base.

As mentioned, a general CBR cycle may be described by four top-level steps:

1. RETRIEVE the most similar cases; During this process, the CB reasoner searches the database to find the most approximate case to the current situation.

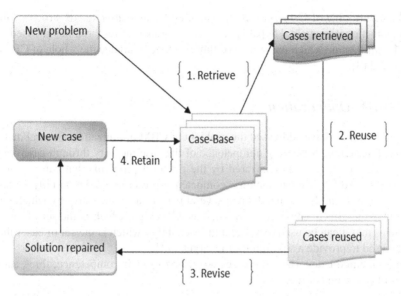

Fig. 1 CBR Cycle

2. REUSE the cases to attempt to solve the problem; This process includes using the retrieved case and adapting it to the new situation. At the end of this process, the reasoner might propose a solution.
3. REVISE the proposed solution if necessary; Since the proposed solution could be inadequate, this process can correct the first proposed solution.
4. RETAIN the new solution as a part of a new case.

   This process enables CBR to learn and create a new solution and a new case that should be added to the case base [2].

   As shown in Figure 1, the CBR cycle starts with the description of a new problem, it retrieves the most similar case or cases. It modifies cases which are retrieved for adapting to the given problem. If the suggested solution is rejected, the CBR system modifies the proposed solution, retains the repaired case and incorporated it into the case base [3]. During the retain step, the CBR can enter into the maintenance phase to increase the speed and the quality of the case base retrieved process especially when the number of cases gets large. As well as the importance of maintenance research.

3   Case Base Maintenance Policies

In this section, we describe some of strategies employed to maintain case memory. Case base maintenance CBM policies differ in the approach they use and the schedule they follow to perform maintenance [9]. Most existing works on CBM are based on one of two general approaches: One branch of research has focused on the partitioning of Case Base (CB) which builds an elaborate CB structure and maintains it continuously [1] [6] [10]. Another branch of research has focused on CBM optimization which uses an algorithm to delete or update the whole of CB [7] [11] [12] [13].

3.1 CBM Optimization

Many researchers have addressed the problem of CBM optimization. Several methods are focused on preserving competence of the case memory through case deletion, where competence is measured by the range of problems that can be satisfactorily solved [7]. We can mention competence-preserving deletion [14]. In this method, Smyth and McKenna defined several performance measures by which one can judge the effectiveness of a CB, such as: Coverage, which is the set of target problems that it can be used to solve, and reachability which is the set of cases that can be used to provide a solution for the target [11].

Their method categorized the cases according to their competence, three categories of cases are considered [14]:

- Pivotal case: If it is reachable by no other case but itself and if its deletion directly reduces the competence of system. Pivotal cases are generally outliers.
- Support cases: They exist in groups. Each support case provides similar coverage to others in group. Deletion of any case in support group does not reduce competence. Deletion of all in group equivalent to deleting pivot
- Auxiliary case: If its coverage set is a subset of the coverage of one of its reachable cases, it does not affect competence at all, its deletion makes no difference.

As a result, the optimal CB can be constructed from all the pivotal cases plus one case from each support group.

Accuracy-Classification Case Memory (ACCM) and Negative Accuracy-Classification Case Memory (NACCM), proposed in [15] where the foundations of these approaches are the Rough Sets Theory. They used the coverage concept which is computed as follows:

Let $T = \{t_1; t_2; ...; t_n\}$ be a training set of instances:

$$Coverage(t_i) = AccurCoef(t_i) \bigoplus ClassCoef(t_i) \qquad (1)$$

Where AccurCoef measure explains if an instance t is an internal region or an outlier region, ClassCoef measure expresses the percentage of cases which can be correctly classified in T, and the $\bigoplus$ operator is the logical sum of both measures.

The main idea of ACCM reduction technique is: Firstly, to maintain all the cases that are outliers, cases with a Coverage = 1.0 value are not removed. This assumption is made because if a case is isolated, there is no other case that can solve it. Secondly, to select cases that are nearest to the outliers and other cases nearby can be used to solve it because their coverage is higher.

NACCM reduction technique is based on ACCM, doing the complementary process. The motivation for this technique is to select a wider range of cases than the ACCM technique [15].

3.2   CBM Partitioning

For the branch of CBM partitioning, we can mention:

The method proposed in [1] partitions cases into clusters where the cases in the same cluster are more similar than cases in other clusters. Clusters can be converted to new case bases, which are smaller in size and when stored distritbuted, can entail simpler maintenance operations. The contents of the new case bases are more focused and easier to retrieve and update. Clustering technique is applicable to CBR because each element in a case base is represented as an individual, and there is a strong notion of a distance between different cases.

Shiu et al. [16] proposed a case-base maintenance methodology based on the idea of transferring knowledge between knowledge containers in a case-based reasoning (CBR) system. A machine-learning technique namely fuzzy decision-tree induction, is used to transform the case knowledge to adaptation knowledge. By learning the more sophisticated fuzzy adaptation knowledge, many of the redundant cases can be removed. This approach is particularly useful when the case base consists of a large number of redundant cases and the retrieval efficiency becomes a real concern of the user. This method consists of four steps: First, an approach of maintaining a feature weights automatically. Second, clustering of cases to identify different concepts in the case base. Third, adaptation rules are mined for each concept. Fourth, a selection strategy based on the coverage and the reachability of cases to select representative cases.

4   Clustering, Ouliers and Internal Cases Detection (COID) Method

The objective of our maintenance approach named COID is to improve the prediction accuracy of CBR system, and at the same time reduce the size of the case memory.

Our COID method, based on Clustering, Ouliers and Internal cases Detection, defines two important type of case which must be exist in the CB (not delete), because they affect the competence of the system (see Figure 2):

- Outlier: is a case isolated, it is reachable by no other case but itself, its deletion directly reduces the competence of system because there is no other case that can solve it.
- Internal case: is one case from a group of similar cases. Each case from this group provides similar coverage to others in group. Deleting any member of this group has no effect on competence since the remaining cases offer the same coverage. However, deleting the entire group is tantamount to deleting an outlier case as

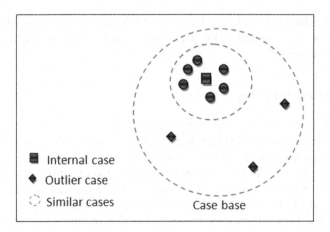

Fig. 2 Case categories in CB

competence is reduced. So, we must keep one case a minimum of one group of similar cases.

Based on these definitions, our COID method reduces the CB by keeping only these two types of cases which influence the competence of CBR. For that, it uses machine learning techniques, by applying clustering method and outliers detection.

The COID's idea is to create multiple, small case bases that are located on different sites. Each small case base contains cases that are closely related to each other. Between different case bases, the cases are farther apart from each other. We try to minimize the intra-cluster distance and to maximize the inter-cluster.distance, so that the degree of association will be strong between members of the same cluster and weak between members of different clusters. This way each cluster describes, in terms of cases collected, the class to which its members belong.

The clustering ensures that each case base is small and it is easier to maintain each one individually. For each small CB, the cases of type outliers and the cases which are near to the center of the group are kept and the rest of cases is removed.

The COID's use of clustering in large CB offers many positive points. In fact, it creates small case bases which are easy to treat, then it detects much more outliers to preserve much competence. Besides, with COID, it is easy to maintain the CB when a new case is learned.

Details of the COID maintaining method consist of two steps (see Figure3):

1. Clustering cases: a large case base is decomposed into groups of closely related cases. So, we obtain independent small CBs.
2. Detection the outliers and the internal cases: outliers and cases which are closed to the center of each cluster are selected and the others cases are deleted.

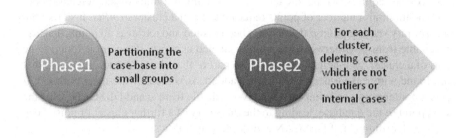

Fig. 3 Two phases of COID

4.1 Clustering Cases

Clustering solves our classification problems. It is applicable to CBR because each case in a case base is independent from other cases, it is represented as an individual element, and there is a different distance between different cases.

Of the many clustering approaches that have been proposed, only some algorithms are suitable for our problem with large number of cases.

We ideally use a method that discovers structure in such data sets and has the following main properties:

- It is automatic; in particular, a principled and intuitive problem formulation, such that the user does not need to set parameters, especially the number of clusters K. In fact, if the user is not a domain expert, it is difficult to choose the best value for K and in our method of maintenance, we need to have the best number of cluster to determinate the groups of similar cases.
- It is easy to be incremental for future learning cases.
- It has not the capability of handling outliers: In our method of maintenance, we need to keep the outliers not to remove them.
- It has the capability of handling noisy cases to guarantee the increase of the classification accuracy.
- It scales up for large case base.

In response, we adopt a typical approach to clustering, density-based clustering method [17]: it shows good performance on large databases and it offers minimal requirements of domain knowledge to determine the input parameters, especially it does not require the user to pre-specify the number of clusters. Moreover, it can detect the noisy cases, so the improve of classification accuracy.

The main idea of density-based approach is to find regions of high density and low density, with high-density regions being separated from low-density regions. These approaches can make it easy to discover arbitrary clusters.

We choose the clustering algorithm DBSCAN (Density-Based Spatial Clustering of Applications with Noise) for its positif points: DBSCAN discovers clusters of arbitrary shape and can distinguish noise, it uses spatial access methods, it is efficient even for large spatial databases [18].

DBScan requires two parameters: *EPS* (Maximum radius of the neighborhood) and the minimum number of points required to form a cluster *MinPts*. It starts with an arbitrary starting point p that has not been visited and retrieves all points density-reachable from p. Otherwise, the point is labeled as noise. If a point is found to be part of a cluster, its e-neighborhood is also part of that cluster. Hence, all points that are found within the e-neighborhood are added, as is their own e-neighborhood. If p is a border point, no points are density-reachable from p and DBSCAN visits the next point of the database, leading to the discovery of a further cluster or noise [19].

The algorithm (1) of DBSCAN is described as follows:

Algorithm 1. Basic DBSCAN Algorithm

1: Arbitrary select a point p.
2: Retrieve all points density-reachable from p w.r.t Eps and MinPts.
3: If p is a core point, a cluster is formed.
4: If p is a border point, no points are density-reachable from p and DBSCAN visits the next point of the database.
5: Continue the process until all of the points have been processed.

4.2   Selecting the Important Cases: Outliers and Internal Cases

Based on a large case base is partitioned, the smaller case bases are built on the basis of clustering result. Each cluster is considered as a small independent case base.

This phase aims at selecting cases which are not to be removed, from each cluster. As mentioned, the selecting cases are the outliers and the internal cases of each cluster: For each small CB, the cases of type outliers and the cases which are near to the center of the group are kept and the rest of cases is removed (see Figure 4).

4.2.1   Internal cases

They are the nearest cases to the center of one group. We calculate the distance between the cluster's center and each case for the same cluster. We choose cases which have the smallest distance.

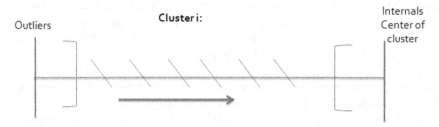

Fig. 4 Selecting cases for one cluster

We consider a case memory which all attributes are supposed to take on numbered values. In this case, we choose to use the Euclidean distance (see Equation 2), which is fast and simple distance's measure:

$$D_{(E_i, M_i)} = (\sum_{j=1}^{n} (E_{ij} - M_{ij})^2)^{1/2} \tag{2}$$

Where, E_i is a case in the cluster C_i, n is the number of cases in C_i and M_i is the cluster's mean of C_i:

$$M_i = \frac{\sum_{i=1}^{n} E_i}{n} \tag{3}$$

4.2.2   Outliers

They are cases that have data values that are very different from the data values for the majority of cases in the data set. They are important because they can change the results of our data analysis [20].

In our case, outliers are important, they present the pivot cases: they affect the competence of the system, they are isolated, there is no other case that can solve them.

For each cluster, COID applies two outliers detection methods to announce the univariate outlier and the multivariate one.

For univariate outlier detection, we choose to use Interquartile Range (IQR), it is robust statistical method, less sensitive to presence of outliers (opposed to Z-score [21]) and quite simple.

IQR defines three quartiles [22]:

- The first quartile (Q1) is the the median of the lower half of the data. One-fourth of the data lies below the first quartile and three-fourths lies above (the 25th percentile)
- The second quartile (Q2) is another name for the median of the entire set of data. Median of data set is the second quartile of data set. (the 50th percentile)
- The third quartile (Q3) is the median of the upper half of the data. Three-fourths of the data lies below the third quartile and one-fourth lies above. (the 75th percentile)

IQR is the difference between Q3 and Q1. It treats any value greater than the 75th percentile plus 1.5 times the interquartile distance, or less than the 25th percentile minus 1.5 times the inter-quartile distance as an outlier.

For multivariate outlier detection, the Mahalanobis distance is a well-known criterion which depends on estimated parameters of the multivariate distribution [23].

Given p-dimensional multivariate sample (cases) x_i $(i = 1; 2...; n)$ the Mahalanobis distance is defined as:

$$MD_i = ((x_i - t)^T C_n^{-1} (x_i - t)^{1/2} \tag{4}$$

where t is the estimated multivariate location and C_n the estimated covariance matrix:

$$C_n = \frac{1}{n-1} \sum_{i=1}^{n} (x_i - \overline{X_n})(x_i - \overline{X_n})^T \tag{5}$$

Accordingly, those observations with a large Mahalanobis distance are indicated as outliers. For multivariate normally distributed data the values are approximately "chi-square" distributed with p degrees of freedom (X_p^2). Multivariate outliers can now simply be defined as observations having a large (squared) Mahalanobis distance. For this purpose, a quantile of the chi-squared distribution (e.g., the 90% quantile) could be considered.

Unfortunately, the "chi-square" plot for multivariate normality is not resistant to the effects of outliers. A few discrepant observations not only affect the mean vector, but also inflates the variance covariance matrix. Thus, the effect of the few wild observations is spread through all the "MD" values. Moreover, this tends to decrease the range of the "MD" values, making it harder to detect extreme ones.

Among many solutions have been proposed for this problem, we choose the using of multivariate trimming procedure to calculate squared distances which are not affected by potential outliers [24].

On each iteration, the trimmed mean $\overline{X}_{trim}$ and trimmed variance covariance matrix, $\overline{C}_{trim}$ are computed from the remaining observations. Then new MD values are computed using the robust mean and covariance matrix:

$$MD_i = ((x_i - \overline{X}_{trim})\overline{C}_{trim}^{-1}(x_i - \overline{X}_{trim})^{1/2} \tag{6}$$

The effect of trimming is that observations with large distances do not contribute to the calculations for the remaining observations.

Consequently of COID method, we build a new reduced case base which contain cases that do not reduce the competence of the system.

5 Experimental Results

In this Section, we try to show the effectiveness of our COID approach as well as the performance of a CBR system.

The aim of our reduction technique is to reduce the case base while maintaining as much as possible the competence of the system. Thus, the sizes of the CBs as regards their competence are compared. The percentage of correct classification has been averaged over stratified ten-fold cross validation runs in front of 1NN.

In order to evaluate the performance rate of COID, we test on real databases obtained from the U.C.I. repository [8]. Details of these databases are presented in Table 1.

We run some well-known reduction methods the Condensed Nearest Neighbor algorithm CNN [25] and Reduced Nearest Neighbor RNN technique [26], on the previous data sets.

Table 1 Description of databases

Database	Ref	#instances	#attributes
Ionosphere	IO	351	34
Mammographic Mass	MM	961	6
Yeast	YT	1484	8
Cloud	CL	1023	10
Concrete-C-strength	CC	1030	9
Vowel	VL	4274	12
Iris	IR	150	4
Glass	GL	214	9

Table 2 Storage Size percentages (S %) of CBR, COID, CNN and RNN methods

Ref	CBR	COID	CNN	RNN
IO	100	6.75	11.3	83.91
MM	100	35.18	64.21	52.26
YT	100	8.22	12.34	14.02
CL	100	48.2	62.3	72.89
CC	100	47.52	62	66.8
VL	100	4.73	31.34	79
IR	100	48.61	17.63	93.33
GL	100	89.62	35.9	93.12

Table 2 shows results for CNN, RNN and COID, it compares the average storage size percentage for the ten runs of these reduction techniques which is the ratio in percentage of the number of cases in the reduced CB that were included in the initial CB, and Table 3 shows the average classification accuracy in percent of the testing data for the ten runs of the previous algorithms.

Looking at Table 2, it can be clearly seen that the reduction rate provided by COID method is notably higher than the one provided by the two policies in the most data-sets, particularly for Ionosphere and Vowel data-sets, it keeps only 7% of the data instances. The same observations is made compared to the Cloud and Mammographic databases, where the obtained CBs are roughly reduced by more than half (see Figure 5). Note that for IRIS and Glass datasets, CNN gives better reudction than COID method and this due to the type of instances and to the fact that these two bases are too small.

From Table 3, we observe that the prediction accuracy obtained using COID method is generally better than CNN and RNN. On the other hand, for the same criterion, results presented by COID are very competitive to those given by CBR with the advantage that COID method reduces significantly retrieval time. This is explained by the fact that our technique reduced CBs' sizes and removed noisy instances (by DBSCAN).

Table 3 Classification accuracy percentages (PCC %) of CBR, COID, CNN and RNN methods

Ref	CBR	COID	CNN	RNN
IO	86.67	86.41	71.9	82.13
MM	85.22	76.52	70.82	78.6
YT	86.16	86.78	83.56	83.92
CL	96.97	94.89	82.92	89.43
CC	99.88	98.44	92.83	86.7
VL	94.35	75.6	81.3	84
IR	95.66	96.49	73	94.23
GL	92.04	92.38	87.45	89.62

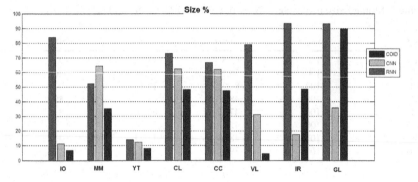

Fig. 5 Comparison of storage Size

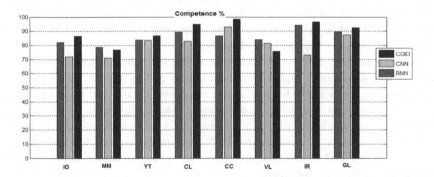

Fig. 6 Comparison of competence

6 Conclusion and Futures Works

In this paper, we have proposed a case-base maintenance method named COID, it is able to maintain the case bases by reducing its size, consequently shortening the case retrieval time. Based on the machine-learning techniques, we have applied clustering analysis and outliers detection methods to reduce the size by preserving maximum competence. The proposed method shows good results in terms of CB reduction size and especially in prediction accuracy.

Our deletion policy can be improved in future works by introducing weighting methods in order to check the reliability of our reduction techniques.

References

1. Yang, Q., Wu, J.: Keep it simple: A case-base maintenance policy based on clustering and information theory. In: Hamilton, H.J. (ed.) Canadian AI 2000. LNCS (LNAI), vol. 1822, pp. 102–114. Springer, Heidelberg (2000)
2. Aamodt, A., Plaza, E.: Case-based reasoning: Foundational issues, methodological variations, and system approaches. Artificial Intelligence Communications 7(1), 39–52 (1994)
3. Leake, D.B.: Case-Based Reasoning: Experiences, Lessons and Future Directions. MIT Press, Cambridge (1996)
4. Sun, Z., Finnie, G.: A unified logical model for cbr-based e-commerce systems. International Journal of Intelligent Systems, 1–28 (2005)
5. Mantaras, L.D., McSherry, D., Bridge, D., Leake, D., Smyth, B., Craw, S., Faltings, B., Maher, M.L., Cox, M.T., Forbus, K., Keane, M., Aamodt, A., Watson, I.: Retrieval, reuse, revision and retention in case-based reasoning. Knowl. Eng. Rev. 20(3), 215–240 (2005)
6. Leake, D.B., Wilson, D.C.: Maintaining case-based reasoners: Dimensions and directions, vol. 17, pp. 196–213 (2001)
7. Haouchine, M.K., Chebel-Morello, B., Zerhouni, N.: Competence-preserving case-deletion strategy for case-base maintenance. In: Similarity and Knowledge Discovery in Case-Based Reasoning Workshop. 9th European Conference on Case-Based Reasoning, ECCBR 2008, pp. 171–184 (2008)
8. Asuncion, A., Newman, D.: UCI machine learning repository (2007)
9. Arshadi, N., Jurisica, I.: Maintaining case-based reasoning systems: A machine learning approach. In: Funk, P., González Calero, P.A. (eds.) ECCBR 2004. LNCS (LNAI), vol. 3155, pp. 17–31. Springer, Heidelberg (2004)
10. Cao, G., Shiu, S., Wang, X.: A fuzzy-rough approach for case base maintenance. LNCS, pp. 118–132. Springer, Heidelberg (2001)
11. Smyth, B., Keane, M.T.: Remembering to forget: A competence-preserving case deletion policy for case-based reasoning systems. In: Proceeding of the 14th International Joint Conference on Artificial Intelligent, pp. 377–382 (1995)
12. Haouchine, M.K., Chebel-Morello, B., Zerhouni, N.: Auto-increment of expertise for failure diagnostic. In: 13th IFAC Symposium on Information Control Problems in Manufacturing, INCOM 2009, Moscou Russie, pp. 367–372 (2009)
13. Haouchine, M.K., Chebel-Morello, B., Zerhouni, N.: Methode de suppression de cas pour une maintenance de base de cas. In: 14 eme Atelier de Raisonnement A Partir de Cas, RaPC 2006, France, pp. 39–50 (2006)
14. Smyth, B.: Case-base maintenance. In: Proceedings of the 11th International Conference on Industrial and Engineering Applications of Artificial Intelligence and Expert Systems IEA/AIE, pp. 507–516 (1998)
15. Salamó, M., Golobardes, E.: Hybrid deletion policies for case base maintenance. In: Proceedings of FLAIRS 2003, pp. 150–154. AAAI Press, Menlo Park (2003)
16. Shiu, S.C.K., Yeung, D.S., Sun, C.H., Wang, X.: Transferring case knowledge to adaptation knowledge: An approach for case-base maintenance. Computational Intelligence 17(2), 295–314 (2001)

17. Bajcsy, P., Ahuja, N.: Location and density-based hierarchical clustering using similarity analysis. IEEE Transactions on Pattern Analysis and Machine Intelligence 20, 1011–1015 (1998)
18. Sander, J., Ester, M., Kriegel, H.P., Xu, X.: Density-based clustering in spatial databases: The algorithm gdbscan and its applications. Data Min. Knowl. Discov. 2(2), 169–194 (1998)
19. Ester, M., Peter Kriegel, H., Jörg, S., Xu, X.: A density-based algorithm for discovering clusters in large spatial databases with noise, pp. 226–231. AAAI Press, Menlo Park (1996)
20. Hautamäki, V., Cherednichenko, S., Kärkkäinen, I., Kinnunen, T., Fränti, P.: Improving k-means by outlier removal. In: Kalviainen, H., Parkkinen, J., Kaarna, A. (eds.) SCIA 2005. LNCS, vol. 3540, pp. 978–987. Springer, Heidelberg (2005)
21. Muenchen, A.R., Hilbe, J.M.: R for Stata Users. Statistics and Computing. Springer, Heidelberg (2010)
22. Bussian, B.M., Härdle, W.: Robust smoothing applied to white noise and single outlier contaminated raman spectra. Appl. Spectrosc. 38(3), 309–313 (1984)
23. Filzmoser, P., Garrett, R.G., Reimann, C.: Multivariate outlier detection in exploration geochemistry. Computers & Geosciences 31, 579–587 (2005)
24. Gnanandesikan, R.: Methods for Statistical Data Analysis of Multivariate Observations, 2nd edn. John Wiley & Sons, New York (1997)
25. Chou, C.H., Kuo, B.H., Chang, F.: The generalized condensed nearest neighbor rule as a data reduction method. In: International Conference on Pattern Recognition, vol. 2, pp. 556–559 (2006)
26. Li, J., Manry, M.T., Yu, C., Wilson, D.R.: Prototype classifier design with pruning. International Journal on Artificial Intelligence Tools 14(1-2), 261–280 (2005)

Congestion and Network Density Adaptive Broadcasting in Mobile Ad Hoc Networks

Shagufta Henna and Thomas Erlebach

Abstract. Flooding is an obligatory technique to broadcast messages within mobile ad hoc networks (MANETs). Simple flooding mechanisms cause the broadcast storm problem and swamp the network with a high number of unnecessary retransmissions, thus resulting in increased packet collisions and contention. This degrades the network performance enormously. There are several broadcasting schemes that adapt to different network conditions of node density, congestion and mobility.

A comprehensive simulation based analysis of the effects of different network conditions on the performance of popular broadcasting schemes can provide an insight to improve their performance by adapting to different network conditions. This paper attempts to provide a simulation based analysis of some widely used broadcasting schemes. In addition, we have proposed two adaptive extensions to a popular protocol known as Scalable Broadcasting Algorithm (SBA) which improve its performance in highly congestive and sparser network scenarios. Simulations of these extensions have shown an excellent reduction in broadcast latency and broadcast cost while improving reachability.

1 Introduction

In ad hoc wireless networks, mobile users are able to communicate with each other in areas where existing infrastructure is inconvenient to use or does not exist at all. These networks do not need any centralized administration or support services. In such networks, each mobile node works not only like a host, but also acts as a router. The applications of ad-hoc wireless networks range from disaster rescue and tactical communication for the military, to interactive conferences where it is hard or

Shagufta Henna
University of Leicester, University Road Leicester, UK
e-mail: sh334@le.ac.uk

Thomas Erlebach
University of Leicester, University Road Leicester, UK
e-mail: t.erlebach@le.ac.uk

Roger Lee (Ed.): SNPD 2010, SCI 295, pp. 53–67, 2010.

expensive to maintain a fixed communication infrastructure. In such a relay-based system, broadcasting is a fundamental communication process used to distribute control packets for various functionalities. It has extensive applications in wireless ad-hoc networks. Several routing protocols such as Ad Hoc On Demand Distance Vector (AODV), Dynamic Source Routing (DSR), Zone Routing Protocol (ZRP), and Location Aided Routing (LAR) typically use broadcasting or a derivation of it to establish their routes, in exchanging topology updates, for sending control or warning messages, or as an efficient mechanisms for reliable multicast in high speed wireless networks.

The simplest broadcasting approach is blind flooding, in which each node is obligated to forward the broadcast packet exactly once in the network without any global information. This causes redundancy that leads to a rather serious problem known as broadcast storm problem [3]. Broadcast storm problem results in high bandwidth consumption and energy use, which devastate network performance enormously.

The main objective of efficient broadcasting techniques is to reduce the redundancy in the network communication while ensuring that the bandwidth and computational overhead is as low as possible. Recent research on different broadcasting techniques reveals that it is indeed possible to decrease the redundancy in the network. This reduces the collisions and contention in the network, which saves the energy of mobile nodes and also results in efficient use of other network resources. Keeping this fact in consideration, several efficient broadcasting schemes have been proposed by different research groups while ensuring improved performance of these protocols. All these schemes perform well when the congestion level is not very high. There are few broadcasting schemes that adapt to some extent to the network congestion, but these schemes lack an efficient mechanism that can distinguish packet losses due to congestion from mobility-induced losses.

In dense networks, same transmission range is shared by multiple nodes. This results in highly connected network which improves the reachability but at the same time can result in a higher number of retransmissions in the network. On the contrary, in sparse networks there is much less shared coverage among nodes; thus due to lack of connectivity some nodes will not receive some of the broadcast packets which results in poor reachability. Different efficient broadcasting and routing schemes have been proposed by different research groups while ensuring the improved performance of these protocols. Some schemes perform well in congestive scenario while suffering in highly dense networks. None of these efficient schemes adapt to both the network conditions of congestion and network density simultaneously.

In this paper we present a simulation based performance analysis of some important broadcasting schemes. We also analyze the effect of congestion and node density on the performance of these broadcasting schemes. Three important metrics, notably reachability (defined as the fraction of nodes that receive the broadcast message), broadcast cost and broadcast speed are used to assess the performance of these protocols. From the simulation results, it is apparent that these factors have significant impact on the performance of all the broadcasting techniques.

We also present two simple adaptive extensions to a well known protocol know as the Scalable Broadcasting algorithm. These extensions adapt according to the network conditions of congestion and node density. Our first extension Congestion Adaptive Scalable Broadcasting Algorithm (CASBA) uses a cross layer mechanism for congestion detection and adapts to it. Our second extension to SBA known as Density Adaptive Broadcasting Algorithm (DASBA) improves the performance of the SBA by adapting to the local node density. Simulations are performed to evaluate the performance of our proposed extensions i.e. CASBA. and DASBA. The simulation results demonstrate that CASBA can achieve better reachability while reducing the broadcast redundancy significantly. Simulation results of DASBA have shown an excellent improvement in broadcast speed and reachability in case of sparse networks and mobile networks.

The paper is organized as follows. Section 2 reviews some widely used broadcasting schemes we have selected for analysis under different network conditions of node density and congestion. Section 3 analyzes the impact of congestion on some existing broadcasting schemes through simulations with ns-2. Section 4 analyzes the effect of node density on performance of broadcasting schemes reviewed in section 2. In section 5 our proposed congestion and node density extension to SBA has been discussed and conclusion is presented in section 6.

2 Broadcasting Protocols

2.1 Simple Flooding

One of the earliest broadcasting schemes in both wireless and wired networks is simple flooding. In simple flooding a source node starts broadcasting to all of its neighbors. Each neighbor rebroadcasts the packet exactly once after receiving it. This process continues until all the nodes in the network have received the broadcast message. Although flooding is simple and easy to implement, it may lead to a serious problem, often known as the broadcast storm problem [3] which is characterized by network bandwidth contention and collisions.

2.2 Scalable Broadcast Algorithm (SBA)

Scalable Broadcast Algorithm (SBA) was proposed by Peng and Lu [7]. In SBA a node does not rebroadcast if all of its neighbors have been covered by previous transmissions. This protocol requires that all the nodes have knowledge of their two-hop neighbors. It schedules the retransmission of the broadcast message after a certain Random Assessment Delay (RAD). During the RAD period, any successive duplicate messages will be discarded, but the information about the nodes covered by such transmissions is recorded. At the end of the RAD period, the broadcast message is retransmitted only if it can cover extra nodes. In SBA, the RAD is distributed uniformly between 0 and a function of the degree of the neighbor with the highest degree divided by the node's own number of neighbors, denoted by Tmax.

Adaptive SBA was proposed in [6]. It adapts the RAD Tmax of SBA to the congestion level. Adaptive SBA assumes that there is a direct relationship between the number of packets received and the congestion in the network.

2.3 Ad Hoc Broadcast Protocol (AHBP)

Ad Hoc Broadcast Protocol (AHBP) [8] uses two-hop hello messages like SBA. However, unlike SBA, it is the sender that decides which nodes should be designated as Broadcast Relay Gateway (BRG). Only BRGs are allowed to rebroadcast a packet. In AHBP, the sender selects the BRGs in such a way that their retransmissions can cover all the two-hop neighbors of the sender. This ensures that all the connected nodes in the network receive the broadcast message, provided that the two-hop neighbor information is accurate. BRG marks these neighbors as already covered and removes them from the list used to choose the next BRGs. AHBP performs well in static networks, but when the mobility of the network increases, the two-hop neighbor information is no longer accurate and thus the reachability of the protocol decreases. To cope with this reduction in reachability, the authors have proposed an extension to AHBP known as AHBP-EX. According to this extension, if a node receives a broadcast message from an unknown neighbor, it will designate itself as a BRG [8].

2.4 Multiple Criteria Broadcasting (MCB)

Reachability, broadcast speed and energy life-time are three important performance objectives of any broadcasting scheme. MCB defines broadcasting problem as a multi-objective problem. MCB is a modification of SBA. It modifies the coverage constraint of the SBA. MCB rebroadcasts a packet if the ratio of covered neighbors is less than a constant threshold α. In MCB, the source node assigns some importance to different broadcast objectives. Neighbor nodes use this priority information along with the neighbor knowledge to rebroadcast the broadcast packet [1].

3 Analysis of Network Conditions on Broadcasting Schemes

Main objective of all the efficient broadcasting schemes is to reduce the number of rebroadcasts while keeping the reachability as high as possible. In this section we analyze the impact of congestion, node density and mobility on the performance of some popular broadcasting schemes. The performance metrics we observed are following:

Broadcast Cost: is defined as the average cost to deliver a broadcast message to all the nodes in the network. Reachability: defines the delivery ratio of a broadcast message and is calculated as Nr/N, Nr represents the number of nodes which have received the flooding message successfully divided by the total number of nodes in the network N. Broadcast Speed: is measured as the inverse of broadcast latency

Table 1 Simulation Parameters

Parameter Name	Parameter Value
Simulation Area	500 m*500 m
Simulation time	1100s
Hello Interval	5s
Jitter Value	0.0001s
Neighbor info timeout	10s
Number of Runs	10

Table 2 Cisco Aironet 350 Card Specifications

Parameter Name	Parameter Value
Receiver Sensitivity	65 dBm
Frequency	2.472GHz
Transmit Power	24 dBm
Channel sensitivity	-78 dBm

which denotes end-to-end delay from the time a source sends a flooding message to the time it has been successfully recieved by nodes in the network.

In order to analyse the impact of congestion, node density and mobility on the performance of the existing broadcasting schemes we have divided our experiments into three trials. Each simulation runs for a simulation time of 1100 seconds. For the first 1000 seconds nodes exchange HELLO messages to populate their neighbour tables. For the last 100 seconds broadcast operations are initiated by the network nodes and messages are disseminated in the network. Simulation parameters are listed in Table 1.

3.1 Congestion Analysis of Existing Broadcasting Schemes

In order to congest the network, we have selected a random node which starts the broadcast process in the network and have varied the flooding rate from 1 packet/s to 90 packets/s. The payload size was kept 64 bytes. All the nodes are moving with a maximum speed of 20 m/s with a pause time of 0 seconds following a random way point movement. The number of nodes we have considered is 70.

Figure 1 illustrates that when messages are disseminated at a rate of 15 packets/sec or less, reachability achieved by ASBA, SBA and MCBCAST remains above 98%. However with an increase in the congestion level, it reduces gradually as larger number of collisions occur in the network. This results in more retransmissions in the network which generate more collisions in the network. AHBP seems less sensitive to the congestion level as it has only few retransmitting nodes. The number of retransmitting nodes of flooding reduces with an increase in congestion which degrades the reachability achieved by flooding during congestion.

From Figure 2 it is clear that the number of retransmitting nodes decreases with an increase in congestion level. This results in fewer nodes which can receive and forward a broadcast message during high congestion level in the network which substantially affects reachability. It is clear from Figure 2 that flooding has 60% retransmitting nodes even in worst congestion scenario because each node is retransmitting in the network, which results in substantial increase in collisions which reduces the number of forwarding nodes. It can be observed from Figure 2 that when the flooding rate is between 15 packets/sec to 45 packets/sec, the number of retransmissions of SBA and ASBA increase. This is due to the fact that a congested network does not deliver the redundant transmissions in the network during RAD period which do not help SBA and ASBA to cancel any redundant transmissions. For a highly congestive scenario, broadcast cost of SBA, ASBA and MCBCAST remains as high as 50%. However AHBP and AHBP-EX have lower broadcast cost than all the other broadcasting protocols for all congestion levels because more retransmission can be avoided due to selected number of BRGs. Figure 3 illustrates that the broadcast speed of all the broadcasting schemes suffers with an increase in congestion level.

3.2   Analysis of Node Density on Existing Broadcasting Schemes

In this study we have varied the density by increasing the number of nodes over a simulation area of 500m*500m. The number of nodes has been varied from 25 to 150 in steps of 25. Each node is moving with a maximum speed of 5 m/sec with a pause time of 0 seconds. We have selected one node randomly to start a broadcast process with a sending rate of 2 packets/s. Figure 5 presents the broadcast cost of each protocol as the node density increases. Figure illustrates that the neighbor knowledge based schemes require fewer retransmitting nodes than flooding. AHBP performs well in denser networks and has a smaller number of retransmitting nodes than all other broadcasting schemes. For a dense network of 150 nodes, AHBP requires less than 50% of the network nodes to rebroadcast.

Figure 4 shows the effects of node density on reachability. Figure illustrates that all the broadcasting schemes are scalable when the network region is dense. The reachability increases almost linearly from the low to medium dense network and reaches 100% for highly dense network. At a high density, network is more connected than at a sparse network. Figure 6 shows the results of the effects of node density on the broadcast speed of all broadcasting schemes. It verifies a strong correlation between the broadcast speed and the node density. It shows that the broadcast speed decreases with an increase in the number of nodes in the network. However AHBP and AHBP-EX have better broadcast speed than other broadcasting protocols because they do not use any RAD for making a rebroadcast decision. Other protocols have better broadcast speed than flooding due to fewer retransmissions.

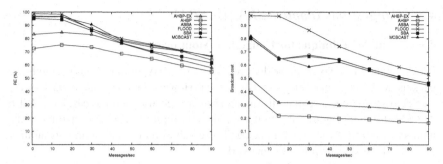

Fig. 1 Reachability vs.Messages/sec **Fig. 2** Broadcast Cost vs.Messages/s

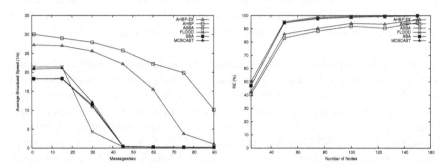

Fig. 3 Broadcast Speed vs.Messages/s **Fig. 4** Reachability vs.# of Nodes

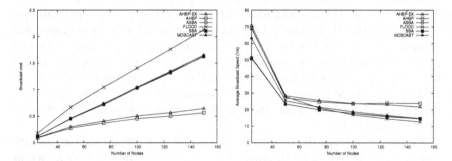

Fig. 5 Broadcast Cost vs.# of Nodes **Fig. 6** Broadcast Speed vs.# of Nodes

4 Proposed Extensions to SBA

In this section we propose two simple extensions which can improve the performance of SBA in highly congestive and sparser network scenarios.

4.1 Congestion-Adaptive SBA (CASBA)

4.1.1 Simple Technique for Congestion Monitoring

Packets at the MAC layer can be discarded due to three types of reasons i.e. duplicate packets, MAC busy and retry exceeded count. However approximately 95% of the packet losses are because of collisions and rest are because of other reasons. In IEEE 802.11 [2] standard short retry limit and long retry limit are used to keep track of the number of retransmissions of a packet. These values represent continuous number of packet losses occurring in the network.

A packet is considered to be correctly received by a node, if it has received power greater than or equal to RxThreshold (Reception Threshold). Any packet having power less than RxThreshold is considered not to be valid and is therefore discarded. Based on the observations made from the MAC layer behaviour, we have used a simple scheme for distinguishing mobility-induced packet losses from the congestion loss.

Long Retry Limit and Short Retry Limit represent continuous number of retransmissions occurring in the network, we have used the packet loss after retry count exceeds in our technique to predict its reason. In our proposed technique a node keeps a record of the received powers of the last ten packets received from the destination. It also records the time when the packet was received by the node from its neighbour. When MAC layer does not receive a CTS/ACK for a certain number of times known as Short Retry Limit or Long Retry Limit, it discards the corresponding packet. In order to know the reason of packet loss, we have taken an average on the received powers of last ten packets and have maintained a variable to keep track if the received power from a certain node is gradually decreasing. If the average of the received powers of last ten packets is less than RxThreshold and the power received is gradually decreasing and the last packet received from the corresponding destination is not older than 3 seconds (to make sure if we are not using an outdated packet), we conclude that packet was lost because of mobility in the network and we set the reason of packet loss as MOBILITY otherwise set it to CONGESTION. However this simple scheme of congestion monitoring is based on the assumption that Physical Layer Convergence Protocol (PLCP) header part of the PLCP frame is always interpretable.

We have monitored the congestion information through simulations (not included) by using ns-2 [5] with different scenarios of mobility and congestion in the network. Simulations results has verified that all packet losses are not due to congestion but some are lost due to mobility.

4.1.2 Cross Layer Model

Different layers of the protocol stack are sensitive to different conditions of the network. For example, MAC layer is considered good at measuring congestion and packet losses in the network, Network layer can calculate the number of neighbors associated with a node and physical layer is good at predicting node movements and node location. Thus we can use the information measured on one layer on the other layer. However layered architecture is not flexible enough to cope with the dynamics of the wireless networks. In [4] it is shown that by exploiting the cross layer design we can improve the performance of the other layers of the protocol stack. Cross layer design differs from the traditional network design, in which each layer of the protocol stack operates independently. Communication between these layers is only possible through the interfaces between them. In the cross layer design, information can be exchanged between non adjacent layers of the protocol stack. By adapting to this information the end-to-end performance of the network can be optimized. In our congestion adaptive extension, we improve the performance of the SBA by using the congestion information directly from the MAC layer.

4.1.3 CASBA

CASBA follows a cross-layer approach where information calculated by the MAC layer is used to improve the performance of the Network layer. CASBA, like SBA, has a self-pruning approach and needs to maintain 2-hop neighbor information. It also requires the ID of the last sender. It works as follows: When a node x receives a broadcast packet from a node w, it excludes all the neighbors of w, N(w), which are common with its own neighbors N(x). Node x only rebroadcasts to the remaining set y=N(x)-N(w). If the set y is not empty, node x schedules the next retransmission after the RAD time. The maximum value of RAD is calculated using the following formula:

$$(dmax/dx) * Tmax \tag{1}$$

where dx is the degree of the node x and dmax is the degree of the node with maximum number of neighbors among the neighbors of x. Tmax and dmax/dx control the value of RAD. Tmax controls the length of the RAD, and dmax/dx makes it likely that nodes with higher degree rebroadcast earlier. A node chooses the RAD for the next transmission uniformly in the interval between 0 and the maximum RAD value. We analyzed through simulations (not included here), it is very critical for broadcasting schemes to wisely tune the value of RAD in order to perform efficiently in a MANET environment. In CASBA, the Tmax value is controlled by the MAC layer. Whenever the MAC layer detects a packet loss because of congestion as classified by the approach discussed above, it sets the value of Tmax to 0.05 seconds to delay retransmissions in the congestive network in order to mitigate congestion. This gives a node more time to cancel any redundant transmissions that would only cause the network to become more congested. However, if the MAC layer reports a packet loss as a result of mobility, a Tmax value of 0.01 seconds is used to speed up the broadcast process.

4.1.4 Performance Analysis

We have used ns-2 packet level simulator to conduct experiments to evaluate the performance behavior of CASBA. For the physical layer configurations we have followed cisco Airnet 350 specifications as shown in Table 2, other simulation parameters are same as given in Table 1.

Effect of the congestion

In the figures below, we have quantified the effect of congestion on four broad-casting schemes of Flooding, CASBA, ASBA and SBA. We have varied the packet origination rate from 1 packets/s to 90 packets/s to make the network congested. Other simulation parameters are same as shown in Table 1. Figure 7 shows the reachability achieved by the four protocols as the network becomes congested. In congested scenario, we have observed that the reachability achieved by CASBA is better than SBA and ASBA and Flooding. Figure 8 shows the number of retransmitting nodes in congested network. With an increase in the congestion level the number of retransmitting nodes reduces when the number of nodes and network area is kept constant. Reachability as shown is Figure 7 shows a direct relationship with the number of retransmitting nodes shown in Figure 8. CASBA has better reachability than ASBA, SBA and Flooding because it has few numbers of retransmitting nodes as the network becomes congested. CASBA minimizes the number of redundant retransmissions in the network as it uses a longer value of RAD to cancel more redundant retransmissions as soon as MAC layer interprets a congestion loss. It improves its performance in congested scenarios and reduces the risk of broadcast storm problem which wastes network resources. However CASBA has poor

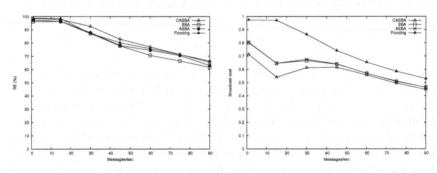

Fig. 7 RE vs.# of Messages   **Fig. 8** Broadcast Cost vs.# of Messages

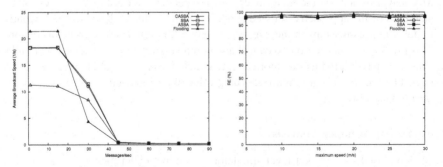

Fig. 9 Broadcast Speed vs.# of Messages   **Fig. 10** RE. vs.Max.Speed

broadcast speed as it adds more delay than ASBA and SBA as soon as it detects a congestion loss as shown in Figure 9. This increase in delay helps CASBA to cancel more redundant transmission which may severely contend the network.

It can be observed from Figure 8 that flooding has the highest broadcast cost among all protocols which leads to broadcast storm problem. Broadcast speed of Flooding as shown in Figure 9 decreases when network becomes more congested due to high number of collisions in the network. It can be observed from Figure 9 that CASBA performs better than flooding in terms of broadcast speed as well in highly congestive scenario.

Effect of Mobility

We used the random way-point model and have varied the maximum speed from 5 m/s to 30 m/s with pause time of 0 seconds. Other simulation parameters are listed in Table 1.

We have measured the performance of all three protocols as the maximum speed increases along x-axis. Figure 10 shows that the reachability of all the three broadcasting schemes is almost the same for all speeds. We have simulated a realistic scenario and have induced significant amount of congestion in the network. The performance of our CASBA protocol is a slightly better than SBA and ASBA. However the number of retransmitting nodes of CASBA is significantly less than SBA, CASBA and Flooding as shown in Figure 11. This is because of the distinction between the congestion and the link failures at the MAC layer. When the MAC layer interprets a retransmission due to congestion, RAD is adapted according to congestion and for link failure it behaves like SBA. Flooding has the highest number of retransmitting nodes as it has no mechanism to control redundant retransmissions in the network. The broadcast speed of CASBA is lower than SBA and ASBA because of its frequent congestion adaptation as shown in Figure 12.

Effect of Node Density

The network considered for the performance analysis of CASBA vs. node density has been varied from 25 nodes to 150 in steps of 25 nodes on a simulation area of 500*500. All the nodes follow a random-way point movement. The packet origination rate of source is 2 packets/sec. Table 1 lists the other simulation parameters. Figure 13 shows that the reachability achieved by all three protocols increases with an increase in density. As the density of the nodes increases, the number of nodes covering a particular area increases which improves reachability of these protocols.

Figure 15 depicts the effect of node density on the broadcast speed. It shows that the broadcast speed is largely affected by the network density, it decreases with an increase in network density. However it can be observed that Flooding has better broadcast speed than CASBA,SBA and ASBA, this is because with an increase in network density number of retransmissions induced by Flooding increases more rapidly than with the other three protocols. This increase in the number of retransmissions makes a broadcast packet reach quickly to the nodes in the network.

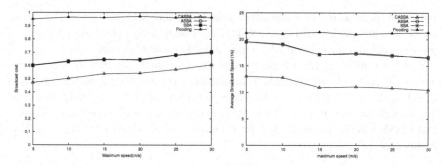

Fig. 11 Broadcast Cost vs.Max.Speed   **Fig. 12** Broadcast Speed vs.Max.Speed

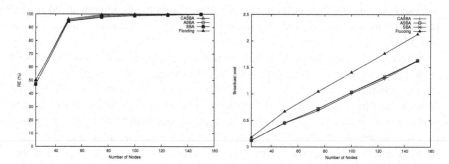

Fig. 13 RE vs.# of Nodes   **Fig. 14** Broadcast Cost vs.# of Nodes

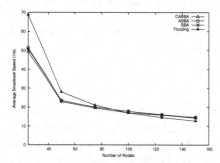

Fig. 15 Broadcast Speed vs.# of Nodes

Although reachability as shown in Figure 13 increases with an increase in node density, a significant amount of congestion in the MANET also arises which results in more retransmissions in the network. This is explained by the fact that more nodes create more contention for accessing the channel. Our CASBA however perform better than SBA and CASBA and adapts to any congestion in the network. Figure 14 shows that it has fewer retransmitting nodes than SBA,ASBA and Flooding which improves its performance in denser networks as well.

From the results it is clear that CASBA performs well in congestive environments or in such environments where packet losses due to congestion or link failure are common. CASBA adjusts the value of RAD by using the information from the MAC layer which differentiates between the packet losses due to congestion and link failures in the network. It cancels more redundant retransmissions during the congestion in the network which alleviates the broadcast storm problem.

4.2 Density Adaptive SBA (DASBA)

In our simple proposed Density adaptive extension of SBA, we have adjusted the value RAD based on the local node density of a node. In case of sparser networks if a node detects a new link with another node, instead of waiting for some RAD period, a node can rebroadcast immediately. In a sparser network there is less chance of congestion and redundant retransmission cancellations as few nodes are involved in the transmission of broadcast message. Adding RAD in such scenario can add only delay and can affect the reachability of the SBA. As an indication of sparser network we have used a value of node density less than or equal to 4 i.e. if a node detects a new link with another node and its current number of neighbors is less than or equal to 4, instead of waiting for some random amount of time it can rebroadcast immediately. An immediate rebroadcast improves the reachability as well as the broadcast speed of SBA. We have used ns-2 packet level simulator to conduct experiments to evaluate the performance behavior of DASBA. For the physical layer configurations we have followed specifications as shown in Table 2, other simulation parameters are same as given in Table 1.

Effect of Node Density

In this section we evaluate the performance of DASBA with different node density values.We used the random way-point model as the mobility model and all the nodes are moving with a maximum speed of 5 m/s with pause time of 0 seconds. Other simulation parameters are listed in Table 1. It is apparent from Figure 16 that reachability achieved by DASBA is better than SBA. It is due to the fact that due to mobility if local node density gets sparser or if there are few number of nodes in the network, instead of waiting for a RAD time a node rebroadcast immediately which improves its reachability. It is clear from the Figure 17 that immediate rebroadcast in DASBA results in an excellent improvement in broadcast speed compared to SBA. However DASBA as shown in Figure 18 has slightly higher broadcast cost than SBA as due to immediate rebroadcast fewer redundant retransmissions can be cancelled. In a sparser network number of redundant retransmissions cancelled are smaller than in a dense network, thus an excellent improvement in broadcast speed and improved reachability seems reasonable at the slight increase in broadcast cost.

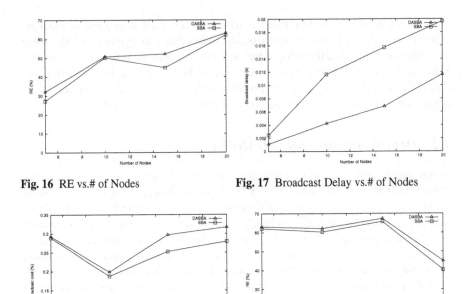

Fig. 16 RE vs.# of Nodes   **Fig. 17** Broadcast Delay vs.# of Nodes

Fig. 18 Broadcast Cost vs.# of Nodes   **Fig. 19** RE vs.maximum speed (m/s)

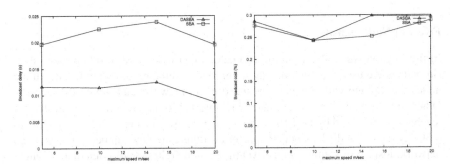

Fig. 20 Broadcast Delay vs. maximum speed **Fig. 21** Broadcast Cost vs. maximum speed
(m/s)   (m/s)

Effect of Mobility

We used the random way-point model and have varied the maximum speed from 5 m/s to 30 m/s with pause time of 0 seconds. Other simulation parameters are listed in Table 1. Maximum number of nodes we have considered for our simulations is 20.

It is apparent from Figure 19 that reachability achieved by DASBA is better than SBA for all mobility values. As a highly mobile network causes a sparser network, an immediate rebroadcast can reach more nodes as compared to a delayed rebroadcast. DASBA in case of high mobility always rebroadcasts earlier than SBA which improves its performance in sparser and mobile scenarios.

It is apparent from Figure 20 that DASBA has much better broadcast speed than SBA due to immediate rebroadcast decision as compared to delayed rebroadcast decision in SBA. However as shown in Figure 21 high mobility will result in more sparser network than without mobility resulting in immediate rebroadcast cancelling only few redundant retransmission which results in higher broadcast cost than SBA.

5   Conclusions

This paper presents a comprehensive simulation based analysis of different broadcasting schemes under different network conditions of congestion and node density. This simulation based analysis provides an insight into different broadcasting schemes and how they work under different conditions of network congestion and node density. We also propose two simple extensions to SBA which improves its performance in sparse and congestive network scenarios. In our congestive extension to SBA known as CASBA, RAD Tmax value is adjusted by the MAC layer according to the estimated reason of the packet loss in the network. The protocol can reduce the broadcast redundancy significantly. It can be adjusted to achieve a balance between broadcast cost and broadcast speed according to the network conditions. We believe that CASBA can result in reduced broadcast redundancy, which saves network resources. Our other simple extension to SBA known as DASBA uses local density information to adjust the value of RAD Tmax which improves its broadcast speed and reachability in mobile and sparser network scenarios.

References

1. Barritt, B.J., Malakooti, B., Guo, Z.: Intelligent multiple-criteria broadcasting in mobile ad-hoc networks. In: Proceedings of 31st IEEE Conference on Local Computer Networks, November 2006, pp. 761–768 (2006)
2. Bianchi, G.: Performance analysis of the IEEE 802.11 distributed coordination function. IEEE Journal of Selected Areas in Communications 18, 535–547 (2000)
3. Ni, S., Tseng, Y., Chen, Y., Sheu, J.: The broadcast storm problem in a mobile ad hoc network. Wireless Networks 8, 153–167 (2002)
4. Srivastava, V., Motani, M.: Cross layer design: A survey and the road ahead. IEEE Communications Magazine 43, 112–119 (2005)
5. UCB/LBNL/VINT. Network simulator-ns-2, http://www.isi.edu/nsnam/ns
6. Williams, B., Camp, T.: Comparison of broadcasting techniques for mobile ad hoc networks. In: Mobihoc Conference Proceedings, June 2002, pp. 194–205 (2002)
7. Peng, W., Lu, X.: On the reduction of broadcast redundancy in mobile ad hoc networks. In: Proceedings of Mobihoc (August 2000)
8. Peng, W., Lu, X.: AHBP: An efficient broadcast protocol for mobile ad hoc networks. Journal of Science and Technology 6, 114–125 (2001)

Design and Performance Evaluation of a Machine Learning-Based Method for Intrusion Detection

Qinglei Zhang, Gongzhu Hu, and Wenying Feng

Abstract. In addition to intrusion prevention, intrusion detection is a critical process for network security. The task of intrusion detection is to identify a network connecting record as representing a normal or abnormal behavior. This is a classification problem that can be addressed using machine learning techniques. Commonly used techniques include supervised learning such as Support Vector Machine classification (SVM) and unsupervised learning such as Clustering with Ant Colony Optimization (ACO). In this paper, we described a new approach that combines SVM and ACO to take advantages of both approaches while overcome their drawbacks. We called the new method *Combining Support Vectors with Ant Colony* (CSVAC). Our experimental results on a benchmark data set show that the new approach was better than or at least comparable with pure SVM and pure ACO on the performance measures.

Keywords: Intrusion detection, machine learning, support vector machine, ant colony optimization.

1 Introduction

As network technology, particularly wireless technology, advances rapidly in recent years, making network secure has become extremely critical more than ever before.

Qinglei Zhang
Department of Computing and Software, Faculty of Engineering,
McMaster University, Hamilton, Ontario, Canada, L8S 4L8
e-mail: zhangq33@mcmaster.ca

Gongzhu Hu
Department of Computer Science, Central Michigan University,
Mt. Pleasant, MI 48859, USA
e-mail: hu1g@cmich.edu

Wenying Feng
Department of Computing & Information Systems, Department of Mathematics,
Trent University, Peterborough, Ontario, Canada K9J 7B8
e-mail: wfeng@trentu.ca

Roger Lee (Ed.): SNPD 2010, SCI 295, pp. 69–83, 2010.

While a number of effective techniques exist for the prevention of attacks, it has been approved over and over again that attacks and intrusions will persist and always be there. Although intrusion prevention is still important, another aspect of network security, intrusion detection [11, 17], is just as important. With trenchant intrusion detection techniques, network systems can make themselves less vulnerable by detecting the attacks and intrusions effectively so the damages can be minimized while keeping normal network activities unaffected.

Two clear options exist for intrusion detection: we can teach the system what all the different types of intruders look like, and then it will be capable of identifying any attackers matching these descriptions in the future, or we can simply show it what normal users look like and then the system will identify any users that do not match this profile. These two alternatives are normally called misuse-based and anomaly-based approaches.

Misuse-based approach [17] involves the training of the detection system based on known attacker profiles. The attributes of previous attacks are used to create attack signatures. During classification, a user's behavior is compared against these signatures, and any matches are considered attacks. A network-connecting activity that does not fit any of the attack profiles is deemed to be legitimate. While this approach minimizes false alarms, it is obvious that new forms of attacks may not fit into any category and will erroneously be classified as normal.

Anomaly-based detection [11] requires the creation of a normal user profile that is used to measure the goodness of subsequent users. Unlike misuse-based detection, a normal-user profile is created instead of attack profiles. Any network connection falling within this range is considered normal, while any users falling out of line with this signature are considered attacks. This approach is more adaptive in that it is far more likely to catch new types of attacks than misuse-based intrusion detection, but it also may lead to a higher false alarm rate. This type of system therefore requires a diverse training set in order to minimize these mistakes.

To overcome the weaknesses of the two approaches, an intelligent learning process (machine learning) can be applied to train the intrusion detection system (IDS) to be able to identify both normal behavior and abnormal behavior including attacks. The very basic machine learning strategies are supervised learning or unsupervised learning. These strategies can apply to intrusion detection. In supervised learning [8], a sample *training data set* with known categories (normal and abnormal) are fed to the system that builds a classification model based on the training data. The system will then use the model to classify new data records of unknown categories as normal or abnormal. In the contrary, unsupervised learning [4] does not rely on a labeled training data set. Instead, it attempts to find structures in the collection of unlabeled data points, such as groups with similar characteristics. Clustering, which is the "process of grouping a set of physical or abstract objects into classes of *similar objects*" [6], is a central technique in unsupervised learning.

Various methods of applying machine learning techniques to intrusion detection have been proposed and developed in the past, most of which are either based on supervised learning or on unsupervised learning, but not on both. There are few, if any, methods that combined both approaches or use one to improve the accuracy/performance of the other.

In this paper, we present an intrusion detection method that builds a classification model using a combination of Support Vector Machine (SVM, supervised learning) and Ant Colony Optimization (ACO, unsupervised learning). The distinguishing feature of the proposed method is to apply both algorithms iteratively in an interweaving way so that the needed data points in the training set is significantly reduced for building a classifier without sacrificing the classification accuracy, and any addition of new training data records will not cause a total re-training of the classifier. That is, the classification accuracy will remain as good as or better than using one approach (supervised or unsupervised learning) alone while the training and model building process will be much faster due to the reduced size of training set.

We will first briefly review the background of machine learning approaches, and then describe the proposed approach called *Combining Support Vectors with Ant Colony* (CSVAC). Experimental results with performance evaluation are then discussed that show that the combined approach did achieve the goal of faster training process and equal or better classification accuracy in comparison to SVM or ACO alone.

2   Review of Machine Learning Methods

In this section, we give a review of the background of several machine learning techniques. The support vector machine and ant colony optimization techniques are the basis of the method presented in this paper. We also include a review of self-organizing maps even it is not directly used in our new approach simply because it is just as suitable as the ant colony optimization technique for clustering, and it will be used in our future study.

2.1   Self-organizing Maps

Self-organizing maps use unsupervised learning in order to cluster data into meaningful classes. It is a type of neural network in which a number of nodes are used to represent neurons in the human brain. These nodes, or neurons, are initially placed randomly in the center of a map, with each node having connections to other nodes. Training then takes place through the use of competitive learning, as training samples are input to the network, and the node with the smallest Euclidean distance from that input is considered to be the winner. In other words, locate j such that:

$$|\mathbf{x} - \mathbf{w}_j| \leq |\mathbf{x} - \mathbf{w}_k| \text{ for all } k \neq j$$

where $\mathbf{x}$ represents the input vector, and $\mathbf{w}_i$ represents the ith node vector. The winning node then moves closer to the input vector using the following learning rule:

$$\Delta \mathbf{w}_j = \alpha(\mathbf{x} - \mathbf{w}_j)$$

where $\Delta \mathbf{w}_j$ represents the change in node j's position, and α is a small learning rate which prevents drastic changes from being made. Along with the winning node, its neighbors also move closer to the input vector, which results in a stretching of the overall node structure. A neighbor is defined by some threshold N, where all prototypes $\mathbf{w}_m$ for which $D(m, j) \leq N$ will be moved along with the winning node. $D(m, j)$ is simply a distance function that measures in terms of distance along the node structures indices [9].

After a sufficient number of training trials are presented to the network, the nodes will eventually stretch out into a structure that fits itself around the data set. Each node will represent a cluster centroid, and new inputs that are presented to the network will be classified based on their Euclidean distance to these nodes. In other words, the winning node is once again determined based on which neuron is closest to the input vector. In this classification phase, nodes are no longer moved, as training has already been completed. The input vector is instead simply clustered with the winning node. This leads to different clusters of data, where n is the number of nodes used in the self-organizing map. In the case of network intrusion detection, these clusters can then be used to classify the input vectors as either being legitimate or illegitimate users [1, 15].

2.2 Support Vector Machines

Support vector machines (SVM) are a classical technique for classification [2, 3]. Like any linear classifier, the goal of a SVM is to find a d-1 dimensional hyperplane to separate two distinct classes in a d dimensional feature space. Support vector machines take this idea one step further, however, as they also stipulate that the decision boundary which provides the maximum margin between classes must be selected. In other words, the technique looks to achieve maximum separation between classes when determining its classification strategy.

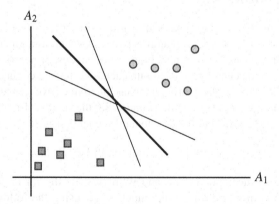

Fig. 1 Decision boundaries of two linearly-separable classes.

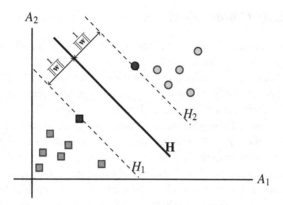

Fig. 2 Maximum-margin hyperplane of support vector machine.

Fig. 1 shows an example of three different decision boundaries for two linearly separable classes in a 2-feature space (A_1–A_2), with only one of them achieving this maximum-margin separation (the thick line). p

The formalization of this idea is as follows. A data point x_i is a vector in a d-dimensional space belonging to either class A or class B. The class label y_i assigned to x_i is

$$y_i = \begin{cases} -1, & x_i \in \text{class } A, \\ 1, & x_i \in \text{class } B. \end{cases}$$

For the hyperplane separating the two classes with the maximum margin, all training data point x_i need to satisfy the constraints

$$\begin{cases} w \cdot x_i - b \geq 1 & \text{for } y_i = 1, \\ w \cdot x_i - b \leq -1 & \text{for } y_i = -1. \end{cases}$$

where w is the normal to H (vector perpendicular to H), b is the distance from H to the origin. This situation is illustrated in Fig. 2. The distance between the two margin hyperplanes H_1 and H_2 is $2/\|w\|$. The data points (darker ones in Fig. 2) on H_1 and H_2 are called *support vectors*.

In order to maximize the total distance, we need to minimize $\|w\|$. The optimization problem then simply becomes to

$$\text{minimize } \|w\|^2/2$$
$$\text{subject to } c_i(wx_i - b) \geq 1 \text{ for all } 1 \leq i \leq n.$$

In the case of non-linearly separable case, the nonlinear network connection records can be classified by using suitable kernel function to transfer them into higher dimension.

2.3   Ant Colony Optimization

Ant colony optimization (ACO) is, as one would expect, a simulation of ants in a colony. Real ants communicate with one another indirectly through the use of pheromones. Initially, ants wander about randomly searching for food. When an ant finds a food source, it returns to the colony and lays down a pheromone trail along the way. Other ants are attracted to these pheromones, and will therefore follow the trail to the food source. These pheromones fade over time, so shorter trails become more attractive to other ants. As a result, short, optimal paths are eventually formed to the closest food sources to the colony.

In the simulation version, "ants" wander about a parameter space in search of an optimal solution for a given problem. In particular, the ant-like agents *pick up* and *drop down* objects in order to achieve an optimal clustering of the objects. The probability $P_p(O_i)$ of picking up object O_i and the probability $P_d(O_i)$ of dropping down the object are calculated based on a swarm similarity function $f(O_i)$ of object O_i:

$$f(O_i) = \sum_{O_j \in Neigh(r)} \left[1 - \frac{d(O_i, O_j)}{\beta} \right]$$

where $Neigh(r)$ is a *neighborhood* of the object O_i, r is the largest distance between O_i and the objects in this neighborhood, $d(O_i, O_j)$ is a distance measure of objects O_i and O_j, and β is a positive parameter controlling the convergence of the algorithm and the number of clusters.

With this probabilistic method, the ants will eventually converge to one optimal path (optimal clustering) which can then be used for classification. In the case of intrusion detection, this technique is used to cluster the training samples of legitimate and illegitimate users. The ants then take new input samples to one of the clusters (i.e. the shortest path) based on the pheromone paths that have been laid, and classification is carried out based on the cluster to which the sample is taken [5, 14].

3   Combination of SVM and ACO

As mentioned in Section 1 (Introduction) that the motivation of our work was to improve the performance of the training process of a supervised learning technique without losing the classification accuracy. One way of achieving this is to combine the supervised learning method SVM and the ACO clustering method. We call the combined algorithm *Combining Support Vectors with Ant Colony* (CSVAC).

3.1   The CSVAC Algorithm

The basic idea of our new CSVAC algorithm is to use SVM to generate hyperplanes for classification and use ACO to dynamically select new data points to add to the active training set for SVM. The two modules interact with each other iteratively and eventually converge to a model that is capable of classify normal and abnormal data records.

First, the original SVM algorithm is modified to become an *active* SVM training process. It repeatedly adjusts the hyperplanes with the maximum margin using the dynamically added data points selected by the ACO clustering algorithm within a neighborhood of the support vectors.

Second, the original ACO algorithm is also modified from clustering all training data to clustering only the training points within a neighborhood of the support vectors. The modification includes selection of neighborhood sizes and the way of assigning cluster labels.

Then, the two modified algorithms are combined into the new algorithm in the training phase of building a classifier, as outlined in Algorithm 1.

Algorithm 1. Training in CSVAC

   Input: A training data set.
   Input: N – number of training iterations.
   Input: RR – detection rate threshold.
   Output: SVM and ACO Classifiers.
   begin
      Normalize the data;
      Let r be the detection rate, initially 0;
      while $r < RR$ **do**
         for $k = 1, \cdots, N$ **do**
            SVM training phase;
            Ant clustering phase;
         end
         Construct classifiers;
         Do testing to update r;
      end
   end

3.2 Design Strategy

An intrusion detection system (IDS) should be able to train a classifier based on the raw audit data records extracted from the network communication and it also should be able to classify new data by using the generated classifier. Furthermore, the training and testing processes of the classifier should be independent with each other. This is a necessary property for real time detection. Therefore, the IDS should be able to perform three basic responsibilities: processing raw data, generating a classifier and classifying new data. In addition, two more independent components are added to the IDS to perform the SVM and ACO functions.

To streamline the data flow between these modules, three types of storages are distributed in the new IDS:

- All candidate data for training*Disk* are stored in the storage called *Disk* where the data type is *point*.
- The training data of each SVM process is named *Cache1*. The data type in Cache1 is called *Lagrangepoint* simply because the optimization process in SVM uses the Lagrange formulation.
- The training data of each ACO clustering phase is named *Cache2* that stores data type *object*.

The main tasks of the five modules are given below.

Raw Data Processing: This is the module directly processing the raw data. The processed data are converted to type *point*, stored in the common storage *Disk*, and used as the input training data by the component SVM and the component AC .

SVM Training: This module implements the SVM training algorithm on the selected data points. The processed data are converted to type *Lagrangepoint* and stored in SVM storage *Cache1*. The support vectors found by this component are passed to the component AC for clustering.

Ant Clustering: The ant colony clustering algorithm is implemented by this module that converts the data points to the type *object* and stored in ant clustering storage *Cache2*. After clustering, the neighbors of those marked objects (support vectors) are fed back to the SVM component as added training data.

Classifier Building: This module builds the CSVAC-based classifier, which is a hybrid of the SVM classifier and the AC classifier. By repeating the processes of SVM training and AC clustering, a classifier is created and stored in the common storage *Disk*, which is used in the testing phase.

Classifier Testing: This is the testing module that tests the classifier stored in *Disk* with a set of new data points. The testing results are the output of the IDS.

3.3   System Components and Implementation

Based on the design strategy discussed above, the implementation of the new IDS consists of the implementation of the different storages and the components.

The responsibility of raw data processing is implemented by the storage *Disk* initialization operation. Storage *Disk* is the common storage used in both of training and testing phases. The entire candidate data for training and the established classifiers for testing are stored here. The class diagrams of storage *Disk* and the data type *point* are shown in Fig. 3.

The responsibility of SVM training is implemented by the component SVM. The selected data for each SVM training process are stored in storage *Cache1*, which is used only by the component SVM. The class diagrams of *Cache1* and the data type *Lagrangepoint* that stored in *Cache1* are shown in Fig. 4.

The marked data for each clustering are stored in storage *Cache2*, which is used only by the component AC. The data type stored in *Cache2* is *object*. The figure for *Cache2* is similar to the one for *Cache1* with some minor differences in their functionalities.

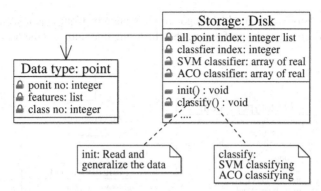

Fig. 3 Class diagram of storage *Disk*

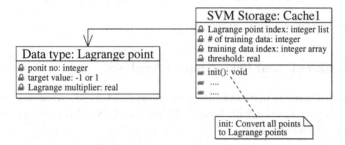

Fig. 4 Class diagram of SVM storage *Cache1*

A method, called Sequential Minimal Optimization (SMO) [13] is used to implement SVM training in the new IDS. The class diagram of the SVM component is shown in Fig. 5.

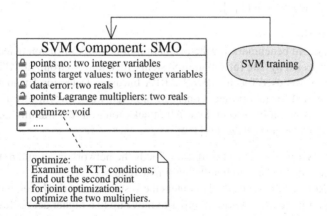

Fig. 5 Class diagram of the SVM component

The responsibility of ant clustering is implemented by Component AC, shown in Fig. 6, that is an ant agent performing the clustering operations in the new IDS.

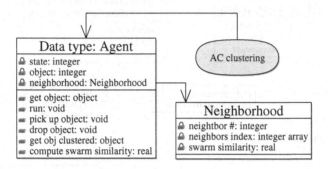

Fig. 6 Class diagram of the AC component

4   Experiments and Performance Analysis

We conducted some experiments using the proposed CSVAC algorithm and compared the its performance with pure SVM and ACO. We also compared with the results of the KDD99 winner.

4.1   Evaluation Data Set

In order to obtain accurate results which can be compared to other similar intrusion detection systems, the KDD99 data set [16] is used for evaluation. This set, which is a version of the DARPA98 data collection, is a standard benchmark set used in the intrusion detection field [7].

DARPA IDS Set
The first standard benchmark for the evaluation of network intrusion detection systems was collected by The Information Systems Technology Group of MIT Lincoln Laboratory. It was used for the 1998 DARPA Intrusion Detection Evaluation. The data set provided for this project consisted of seven weeks worth of training data and two weeks worth of testing data. All attacks found in the training and testing sets fall into four major categories [10]:

1. Denial of Service (DoS): The attacker floods the network in an attempt to make some resource too busy to handle legitimate requests.
2. User to Root (U2R): The attacker enters the system as a normal user, but exploits a vulnerability in the network in order to move up to a higher access level.
3. Remote to Local (R2L): An attacker gains local access to a machine from a remote location without actually having an account on that machine.
4. Probing: An attacker scopes the network to find any weaknesses that can be used for future attacks.

KDD99 Data Set

The KDD99 data set [16], which uses a version of the 1998 DARPA set, has become the standard benchmark data set for intrusion detection systems. The set consists of a large number of network connections, both normal and attacks. Attacks belong to the four main categories discussed above, but can also be broken down in smaller subcategories. Each connection consists of 41 features which can be broken down into the following four categories [2]:

1. Basic features common to all network connections, such as duration, the protocol used, the size of the packet.
2. Traffic features based on a 2-second time window that cannot be measured using a static snapshot of connection, such as DoS and probing attacks.
3. Host-based traffic features that need a much longer time period to collect such as probing.
4. Connection-based content features based on domain knowledge.

Each connection in the KDD99 data set is labeled as either normal or as one of the four attack types described earlier. There are 23 specific types of attacks within these four categories.

The network connection records in KDD99 thus can be classified into 5 classes. The distribution of connection data classes in the 10% KDD99 data set is shown in Table 1. It is a standard data set for the evaluation of IDS and will provide the training and testing data for our experiments.

Table 1 Class distributions of 10% KDD99 data

Class	Number of Data Records
Normal	97277
DoS	391458
U2R	52
R2L	1126
Probe	4107
Total	494021

We used a small portion of the available KDD99 data in our experiment: 390 records for training and 3052 records for testing, and the number of normal and abnormal records used are almost the same. The experiment was repeated five times with different data records selected from the KDD99 data for training.

4.2 Comparison of SVM, ACO, and CSVAC

The proposed CSVAC algorithms was compared with the original SVM and ACO algorithms. Different training data sets were used for the five experiments. Four

Table 2 Comparison of performance measures

Measure	Algorithm		
	SVM	ACO	CSVAC
Training Time (s)	4.231	5.645	3.388
Detection Rate (%)	66.702	80.100	78.180
False Positive (%)	5.536	2.846	2.776
False Negative (%)	21.900	0.360	0.300

measures were calculated from the five observations: training time, detection rate, false positive, and false negative. The mean values of these measures are shown in Table 2.

The visual comparison of the performance measures, except the training time (that is on a different scale), is shown in Fig. 7.

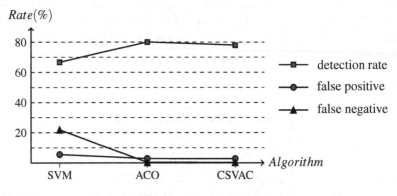

Fig. 7 Performance omparison of the three algorithms.

We also did t-tests on the difference of the mean values with 90% confidence level. The degree of freedom in the t-test for the 3 algorithms and 5 observations is $df = 3 \times (5-1) = 12$. The confidence intervals of the t-scores are given in Table 3. In the table, α_i is the *effect* value of algorithm i for $i = 1, 2, 3$, calculated as

$$\alpha_i = \mu_i - \mu$$

where μ_i is the mean value of the measure of algorithm i and μ is the overall mean of the measure across all the three algorithms. The subscripts 1, 2, and 3 represent SVM, ACO, and CSVAC, respectively.

If the confidence interval does not include zero, we can reject the hypothesis that the difference $\alpha_i - \alpha_j$ has a 0 mean value. From the t-test results, we can conclude with 90% of confidence that the proposed new CSVAC algorithm is better than

Table 3 Confidence intervals of *t*-tests

Measure	Confidence Interval	
	$\alpha_1 - \alpha_3$	$\alpha_2 - \alpha_3$
Training Time (s)		(2.07, 2.44)
Detection Rate (%)	(3.31, 19.65)	(-6.38, 10.07)
False Positive (%)	(2.27, 2.85)	(-0.16, 0,42)
False Negative (%)	(19.43, 23.77)	(-2.11, 2,23)

ACO on training time, better than SVM on detection rate, false positive rate and false negative rate, and comparable with ACO on these rate measures.

4.3 Comparison with KDD99 Winner

We also compared the performance results of our algorithm with the performance results of an algorithm which is the KDD99 winner [12]. The training data of the KDD99 winner were about 20 percents of the standard 10% KDD99 data set and the testing data of the KDD99 winner was the entire 10% KDD99 data set. The testing set is about 50 times of the training data set. The distributions of all classes in the testing set are approximately same as the relevant distributions in the entire 10% KDD99 data set.

Table 4 shows the comparison of CSVAC and the KDD99 winner's results.

Table 4 Comparison of CSVAC and KDD99 winner

Measure	Algorithm	
	CSVAC	KDD99 Winner
Detection Rate (%)	94.86	95.30
False Positive (%)	6.01	4.25
False Negative (%)	1.00	3.00

The comparison results indicate that the new algorithm is comparable to the KDD winner in terms of average detection rate as well as false positive rate and even better than the KDD winner in term of false negative.

5 Conclusion

Machine learning techniques, both supervised and unsupervised, are ideal tools for network intrusion detection. In this paper, we have discussed several primary learning approaches, particularly the Support Vector Machine (SVM) and Ant Colony Optimization (ACO), for intrusion detection, and presented a new method (CSVAC) that combines SVM and ACO to achieve better performance for faster training time by reducing the training data set while allowing dynamically added training data points.

The experimental results indicated that CSVAC performs better than pure SVM in terms of higher average detection rate and lower rates of both false negative and false positive; and it is better than pure ACO in terms of less training time with comparable detection rate and false alarm rates. The performance of the CSVAC algorithm on the experimental data is also comparable to the KDD99 winner.

Differing from other commonly used classification approaches (such as decision tree, k nearest-neighbor, etc.), there is no need for classifier re-training in the proposed CSVAC algorithm for new training data arriving to the system.

We are currently conducting more tests on the algorithm, and working on other machine learning methods, such as neural networks, that can be added to the system as new modules.

References

1. Bivens, A., Embrechts, M., Palagiri, C., Smith, R., Szymanski, B.: Network-based intrusion detection using neural networks. In: Proceedings of the Artificial Neural Networks in Engineering, pp. 10–13. ASME Press (2002)
2. Burges, C.J.C.: A tutorial on support vector machines for pattern recognition. Data Mining and Knowledge Discovery 2(2), 121–167 (1998)
3. Duan, D., Chen, S., Yang, W.: Intrusion detection system based on support vector machine active learning. Computer Engineering 33(1), 153–155 (2007)
4. Duda, R.O., Hart, P.E., Stork, D.G.: Pattern Classification. In: Unsupervised Learning and Clustring, 2nd edn., ch. 10, Wiley, Chichester (2001)
5. Gao, H.H., Yang, H.H., Wang, X.Y.: Ant colony optimization based network intrusion feature selection and detection. In: Proceedings of International Conference on Machine Learning and Cybernetics, vol. 5, pp. 3871–3875 (2005)
6. Han, J., Kamber, M.: Data Mining Concepts and techniques, 2nd edn. Morgan Kaufmann, San Francisco (2006)
7. Kayacik, H.G., Zincir-Heywood, A.N., Heywood, M.I.: Selecting features for intrusion detection: A feature relevance analysis on KDD 99 intrusion detection datasets. In: Proceedings of the Third Annual Conference on Privacy, Security and Trust (2005)
8. Kotsiantis, S.B.: Supervised machine learning: A review of classification techniques. Informatica 31(3), 249–268 (2007)
9. Lichodzijewski, P., Zincir-Heywood, A.N., Heywood, M.I.: Host-based intrusion detection using self-organizing maps. In: Proceedings of the IEEE International Joint Conference on Neural Networks, vol. 2, pp. 1714–1719. IEEE, Los Alamitos (2002)
10. Lincoln Laboratory, MIT: Intrusion detection attacks database (2009), http://www.ll.mit.edu/mission/communications/ist/corpora/ideval/docs/attackDB.html
11. Papaleo, G.: Wireless network intrusion detection system: Implementation and architectural issues. Ph.D. thesis, University of Geneva (2006)
12. Pfahringer, B.: Wining the KDD99 Classification Cup: Bagged boosting. SIGKDD Explorations Newsletter 1(2), 65–66 (2000)

13. Platt, J.C.: Fast training of support vector machines using sequential minimal optimization, pp. 185–208 (1999)
14. Ramos, V., Abraham, A.: ANTIDS: Self-organized ant-based clustering model for intrusion detection system. In: Proceedings of the 4th IEEE International Workshop on Soft Computing as Transdisciplinary Science and Technology, pp. 977–986. Springer, Heidelberg (2005)
15. Rhodes, B.C., Mahaffey, J.A., Cannady, J.D.: Multiple self-organizing maps for intrusion detection. In: Proceedings of the 23rd National Information Systems Security Conference (2000)
16. UCI KDD Archive: KDD Cup 1999 data (1999), http://kdd.ics.uci.edu/databases/kddcup99/kddcup99.html
17. Zhang, Y., Lee, W.: Intrusion detection in wireless ad hoc networks. In: Proceedings of the Sixth Annual International Conference on Mobile Computing and Networking, pp. 275–283 (2000)

9. [...] ECG-based [...] verbose emphasis [...] image-sequence [...] numerical, [...] *apping techniques* 155–206 (2002)

10. Kamei, Y., Monden, A.: ANGLDS: Self-organized map-based clustering model for [...] analysis system. In: Proceedings of the 16 IEEE International Workshop on Social Computing & Transdisciplinary Science and Technology, pp. 979–986. Springer [...] Heidelberg (2012)

11. [...] Bharati, B., Guo, Carmen, J.D.: Principle self-organizing map in [...] In Proceedings [...] for Neuroscience of the Cognition based Information Systems Studies [...] Conference, [...]

12. IEEE KDD and [...] ICML Copp 2006, data mining [...] techniques 685–693, Springer-Verlag, [...] New York, pp. 1–25. Reading 258–1[...]

13. Zhang, Y., Li, [...] Attention Attention in similarity-labor network. Int. [...] and Principle-aware International Conference of Mobile Computing and Science [...] [...] 26, 28–75 (2013)

Evolution of Competing Strategies in a Threshold Model for Task Allocation

Harry Goldingay and Jort van Mourik

Abstract. A nature inspired decentralised multi-agent algorithm is proposed to solve a problem of distributed task allocation in which cities produce and store batches of different mail types. Agents must collect and process the mail batches, without global knowledge of their environment or communication between agents. The problem is constrained so that agents are penalised for switching mail types. When an agent process a mail batch of different type to the previous one, it must undergo a change-over, with repeated change-overs rendering the agent inactive. The efficiency (average amount of mail retrieved), and the flexibility (ability of the agents to react to changes in the environment) are investigated both in static and dynamic environments and with respect to sudden changes. New rules for mail selection and specialisation are proposed and are shown to exhibit improved efficiency and flexibility compared to existing ones. We employ a evolutionary algorithm which allows the various rules to evolve and compete. Apart from obtaining optimised parameters for the various rules for any environment, we also observe extinction and speciation.

1 Introduction

The performance of any distributed system (e.g. distributed heterogeneous computing systems [10], mobile sensor networks [11]) is dependent on the coordination of disparate sub-systems to fulfil the overall goal of the system. The problem can be

Harry Goldingay
Non-linearity and Complexity Research Group (NCRG), Aston University, Aston Triangle, Birmingham, B4 7ET, UK
e-mail: goldinhj@aston.ac.uk

Jort van Mourik
Non-linearity and Complexity Research Group (NCRG), Aston University, Aston Triangle, Birmingham, B4 7ET, UK
e-mail: vanmourj@aston.ac.uk

Roger Lee (Ed.): SNPD 2010, SCI 295, pp. 85–98, 2010.
springerlink.com

seen as one of efficiently deploying a finite set of resources in order to complete
a distributed set of sub-tasks, where these sub-tasks further this overall goal. It is
clear that, in theory, the best method for coordinating these resources must be centralised. A central controller can, at minimum, issue commands causing resources
to be deployed as if according to the optimal decentralised behaviour.

Large systems, however, pose problems for centralised approaches [7] and, as it
is likely that many practical applications will require such large scale systems [14],
decentralised approaches must be investigated. In a decentralised approach, a set of
autonomous, decision making, agents control the behaviour of sub-systems with the
aim of coordinating to provide good global behaviour. A simple method to promote
coordination would be to allow all agents to communicate freely, however n agents
are communicating with each other involves a total of $O(n^2)$ communications [14].
If communication is costly, this is hardly an ideal approach and so finding good
decentralised methods for distributed task allocation, which require a minimum of
communication, is an important problem.

The behaviour of social insects has been a good source of inspiration in the design
of multi-agent-systems (MAS) [2]. In particular, we are interested in algorithms using the principle of "stigmergy", introduced by Grassé [9]. Stigmergic mechanisms
are ones in which agents coordinate themselves using the environment rather than
through direct communication. Behaviours, which are triggered by observation of
the environment, also cause environmental change, thus modifying the behaviour
of other agents. This coordination without communication makes stigmergy a very
attractive paradigm when designing MAS.

The task allocation algorithms we investigate in this paper are based on the response threshold model of task allocation in social insects. Introduced by Bonabeau
et al. [5] it assumes that tasks an individual performs can be broken down into a
finite set of types. The individual has a set of thresholds θ, one for each type of
task, which indicate the agent's willingness to perform each task-type. Instances of
tasks have some environmental stimulus which indicate the priority of the task. The
individual compares the stimulus s of a task with the corresponding threshold θ
to determine the probability of uptake, which should be high for $s \gg \theta$, low for
$s \ll \theta$, zero for $s = 0$ (no demand for the task), and $\frac{1}{2}$ for $s = \theta$, and which is
defined by a *threshold function* $\Theta(s, \theta)$.

Theraulaz et al. [16], proposed a modified version of the model, the variable response threshold model, to account for the genesis of specialisation. In this version.
individuals change their thresholds through self-reinforcement depending on which
task they perform and become specialised in tasks which they carry out frequently.
In this paper, we consider an algorithm based on a discretised version of this model,
where increases in thresholds at times of inactivity are discounted. An agent with
thresholds $\theta = (\theta_1, ... \theta_n)$ will, upon completion of a task of type i, update its thresholds to $\theta' = (u(\theta_1, i), ..., u(\theta_n, i))$ where

$$u(\theta_i, j) \begin{cases} < \theta_i & \text{if } i = j, \\ > \theta_j & \text{otherwise,} \end{cases} \tag{1}$$

with the θ_i restricted to $[\theta_{min}, \theta_{max}]$. u is known as an update rule, and defines agents' task specialisation behaviour.

We introduce a set of new update rules and analyse their behaviour when applied to a general model of distributed task allocation, the "mail processing" problem. We use these rules as "species" in an evolution strategies algorithm to determine the best rule in a given circumstance and to optimise the parameters of these rules. These are statistically analysed from a population dynamics perspective. The flexibility of the system is also tested in a dynamic environment in which task production probabilities are continuously varied, and with respect to a catastrophic breakdown in which all agents specialised in a particular task cease functioning.

The rest of of the paper is organised as follows:
In section 2, we introduce the model and the various strategies to solve it. In section 3, we present and discuss the numerical results. Finally, in section 4, we summarise our main findings, discuss the limitations of the current setting, and give an outlook to future work.

2 The Model

In order to analyse the behaviour and performance of a task allocation algorithm, we need a problem upon which to test it. We choose the the mail processing problem introduced by Bonabeau et al. [4] , developed by Price and Tiňo [12, 13] and modified by Goldingay and van Mourik [8] in order to allow reasonable simulation of large-scale systems.

In this problem, there is a set of N_c cities, each of which is capable of producing and storing one batch each of N_m mail types. The cities are served by a set of N_a agents, each of which has an associated mail processing centre. Agents must travel to a city, choose a batch of mail and take this batch to their processing centre, before repeating this process. There are, however, differences between the mail types and each agent a, has a mail-specialisation σ_a, indicative of the mail type its processing centre can efficiently process. Processing mail of this type takes the centre a fixed time t_p, while the centre must undergo a *changeover* in order to process mail of type $m \neq \sigma_a$, taking a total time $t_c > t_p$. After the changeover, the centre is specialised to deal efficiently with mail type m.

Each centre also has a mail queue in order to buffer the immediate effects of changeovers. This queue is capable of holding up to L_q batches of mail and, while there is space in the queue, the agent continues to collect mail (i.e. remains *active*). When a processing centre finishes processing a batch of mail, it will immediately start processing the next batch in its queue (i.e. the batch which was collected the longest time ago), thus freeing a space in the queue. As centres must process mail in the order in which it was collected, σ_a denotes the mail type last taken by agent a and will be the specialisation of the centre when it comes to process the next collected piece of mail.

In order to simulate this system, we discretise time into steps of the amount of time it takes an agent to visit a city and return with mail. This allows us to define our measure of an algorithms performance, i.e. the *efficiency*, as the average amount of mail processed per agent per time step. During each time step the following happens:

1. Each city which is missing a batch of mail of any type produces a new batch of mail of this type.
2. Each active agent chooses and visits a city.
3. Each city randomly determines the order in which its visiting agents are allowed to act.
4. Each agent examines the available mail at the city in a random order, selecting or rejecting each batch individually until either one is selected or all are rejected.
5. Each agent returns to its processing centres and deposits any mail it has collected in the queue.
6. Each processing centre either processes the next piece of mail in its queue, or continues its changeover.

Further, two different types of environment are considered:

- A static environment, in which a city automatically produces a new batch of mail for every mail type that has been taken at the end of each time step.
- A dynamic environment, in which the probability of batches of mail type being produced varies over time.

The dynamic environment is designed to test the flexibility of the system with respect to continuous changes. The probability of taken mail batches being replaced varies in a sinusoidal fashion (the exact form is not critical), and all mail types have the same wavelength (e.g. to mimic seasonal variations), but a different phase. All mail types have periods of both high and low production. The probability of creating a taken batch of type m at the end of cycle t is given by:

$$\pi_m(t) = \begin{cases} 1, & \text{static} \\ \frac{1}{2}[1 + \sin(\frac{t2\pi}{\xi} - \frac{m2}{\pi}N_m)], & \text{dynamic} \end{cases} \tag{2}$$

where ξ is the wavelength.

In addition to environmental changes, we also investigate the robustness of the system under abrupt changes. To this purpose, we consider the case where all agents specialised in a particular mail type are suddenly removed, and monitor the ability of the system to adapt quickly to the change. This represents one of the most catastrophic failures in terms of instantaneous loss of efficiency, and should give an fair indication of the general robustness.

Note that the definition of this problem as "mail processing" is completely arbitrary. Cities are merely localised task-sites, and the mail types are just different types of tasks. The processing centres and their behaviour can be seen as a generic way to introduce a cost into switching task type. In fact Campos et al. [6] study an almost identical problem but describe it as one of truck painting.

The problem of efficiency maximisation can also be seen as minimisation of loss of efficiency, and we have identified the causes for agents to fail take up mail. For an agent with specialisation σ_a, the efficiency loss sources are:

(ℓ.1) The agent is inactive due to a full queue.

(ℓ.2) The visited city has mail of type σ_a, but the agent rejects all mail.

(ℓ.3) The visited city has some mail, but none of type σ_a, and the agent rejects all mail.

(ℓ.4) The visited city has no available mail at the time of the agent's action.

While a fair proportion of the loss sources is determined by the agent's city choice, an agent still needs an efficient method of decision making concerning the take up of mail once it has arrived at a city. It has been shown that a model known as the variable threshold model is a good mechanism for controlling this tradeoff[12, 13].

2.1 Threshold Based Rules

In order to apply this model to our problem, each mail type represents a different type of task. A batch of mail is seen as an instance of a task and the stimulus is taken to be the amount of time it has been waiting at a city. When a batch of mail is created it is assigned a waiting time of 1 and this time increases by 1 every time step. Each city c, therefore, has a set of waiting times $\mathbf{w}_c = (w_{c,1}, \dots w_{c,N_m})$ where $w_{c,m}$ is the waiting time of the mth mail type.

Each agent a is given a set of thresholds $\theta_a = (\theta_{a,1}, \dots \theta_{a,N_m})$, where $\theta_{a,m}$ is the agent's threshold for taking mail type m. Upon encountering a task of type m with stimulus w, the probability of an agent a accepting is determined by its threshold $\theta_{a,m}$ and its *threshold function* $\Theta(\theta_{a,m}, w)$. Here, we have opted for the exponential threshold function:

$$\Theta(\theta, w) = \begin{cases} 0 & \text{if } w = 0, \\ \frac{w^\lambda}{w^\lambda + \theta^\lambda} & \text{otherwise,} \end{cases} \tag{3}$$

where λ is some appropriately chosen positive exponent.

For a given threshold function, the efficiency of an agent critically depends on its thresholds, while its flexibility to adapt to new situations critically depends on its ability to modify its thresholds or *update rule*. The main goals of this paper are to investigate what kind of update rule is best suited to the problem at hand, and to investigate whether optimal update rules can be found autonomously by competition between the agents. Therefore, we compare the performance of some existing and some newly introduced update rules. We now proceed with a short overview.

The **Variable Response Threshold (VRT)** rule, proposed in [16], was applied to the current problem in [13], [12]. The change in threshold over a period of time t, is given by: $\Delta\theta_m = -\varepsilon\Delta t_m + \psi(t - \Delta t_m)$, where Δt_m is the time spent performing task m, ε, ψ are positive constants, and θ_m is restricted to the interval $[\theta_{min}, \theta_{max}]$. For the current model, time is discretised and thresholds only change when a task is performed. Over a time step, Δt_m is 1 if mail type m was taken and 0 otherwise, and the rule becomes

$$u(\theta_m, i) = \begin{cases} \theta_m - \varepsilon & \text{if } i = m, \\ \theta_m + \psi & \text{otherwise.} \end{cases} \tag{4}$$

The **Switch-Over (SO)** rule fully specialises in the last taken mail type, and fully de-specialises in all other:

$$u(\theta_m, i) = \begin{cases} \theta_{min} & \text{if } i = m, \\ \theta_{max} & \text{otherwise.} \end{cases} \tag{5}$$

In fact, the switch-over rule is an extreme case of the VRT rule with ε, $\psi \geq \theta_{max} - \theta_{min}$. As such values are not in the spirit of VRT, we consider SO as a separate case.

The **Distance Halving (DH)** rule, keeps the thresholds in the appropriate range by halving the Euclidean distance between the current threshold and the appropriate limit.

$$u(\theta_m, i) = \begin{cases} \frac{\theta_m + \theta_{min}}{2} & \text{if } i = m, \\ \frac{\theta_m + \theta_{max}}{2} & \text{otherwise.} \end{cases} \tag{6}$$

Note that DH takes several time steps to fully specialise, but effectively de-specialises in a single time step.

The **(Modified) Hyperbolic Tangent ((m)tanh)** rule is introduced in a similar spirit to the VRT rule, allowing continuous variation of the thresholds without artificial limits. The thresholds are a function of some hidden variables h_m:

$$\theta_m = \theta_{min} + (\theta_{max} - \theta_{min}) \frac{1}{2} (1 + \tanh(h_m)), \tag{7}$$

Note that any sigmoid function could replace $\tanh()$. The update rule works on the h_m similarly to the VRT:

$$u'(h_m, i) = \begin{cases} h_m - \alpha & \text{if } i = m \text{ and } h_m \leq 0, \\ \eta\, h_m - \alpha & \text{if } i = m \text{ and } h_m > 0, \\ h_m + \beta & \text{if } i \neq m \text{ and } h_m \geq 0, \\ \eta\, h_m + \beta & \text{otherwise.} \end{cases} \tag{8}$$

with constants α, $\beta > 0$. We have introduced a re-specialisation coefficient $\eta \in [-1, 1]$ to be able to avoid saturation (large values of h_m producing insignificant changes to the θ_m). This interpolates between the tanh rule for $\eta = 1$, and the SO rule for $\eta \simeq -1$.

2.2 Evolution Strategies

As we wish to study the absolute performance of our update rules in a manner unbiased by our original parameter choices, we optimise our parameter sets. Evolutionary algorithms seem a natural way of optimising the model, as it is inspired by the behaviour of social insects, and have been used before to good effect in similar settings (e.g. genetic algorithms in [3], [6]). We choose to use an evolution strategies (ES) algorithm [1] as its real valued encoding fits well with our problem. ES also allows for inter-update-rule competition, which enables us to find an optimal rule-set while we optimise the parameters rather than optimising each rule-set and choosing the best one. It is also possible that a population of different update rules can outperform a homogeneous population.

To optimise a function f (a measure of solution quality, not necessarily a mathematical function), ES uses a population of individuals with variables $(\mathbf{x}, \sigma, F)$. Here the elements of $\mathbf{x}$, referred to as the *object parameters*, are the inputs into the function we wish to optimise. σ are known as *strategy parameters* and control the mutation of $\mathbf{x}$. Mutation is also applied to the strategy parameters, allowing them to self-adapt to the structure of the fitness space. The *fitness*, $F_m = f(\mathbf{x}_m)$, is a measure of the quality of $\mathbf{x}$ for optimising our function f.

An ES algorithm starts generation g with a set of μ *parents*, $\mathscr{P}(g)$ and proceeds to create a set of ℓ offspring, $\mathscr{O}(g)$. Each offspring agent is selected by choosing ρ (called the mixing number) parents at random from $\mathscr{P}(g)$ and *recombining* them into a single individual. This new individual is then *mutated* according their strategy parameters. These populations undergo *selection*, keeping the fittest μ individuals as a new generation of parents.

In this paper we will not discuss on recombination schemes as we take $\rho = 1$ and, as such, when a parent $(\mathbf{x}, \sigma, F)$ is chosen to produce offspring $(\check{\mathbf{x}}, \check{\sigma}, \check{F})$, we copy the parent's variables. This initial offspring is then mutated, component-wise for each dimension i, and its fitness is evaluated according to the following procedure:

1. $\check{\sigma}_i = \sigma_i \cdot e^{\xi_i}$
2. $\check{x}_i = x_i + \check{\sigma}_i \cdot z_i$
3. $\check{F} = f(\check{\mathbf{x}})$

where $\xi_i \sim N(0, \frac{1}{\sqrt{d}})$ and $z_i \sim N(0,1)$ are independent random numbers and d is the number of dimensions of our object parameter $\check{x}$.

Once a complete set of ℓ offspring has been created, the population undergoes selection during which those μ individuals with the highest fitness from the current set of offspring and parents ($\mathscr{P}(g) \cup \mathscr{O}(g)$) are deterministically selected, and used as the new generation of parent agents $\mathscr{P}(g+1)$. The repeated iteration of this process comprises an ES algorithm.

There are two peculiarities to note in our use of ES. Firstly, we are interested in looking at a range of update rules or "species" (possibly with different parameters spaces) within the same evolving population. As such, at initialisation we define each agent's object parameters to be a random update rule and its necessary parameters. This causes problems within ES as neither mutation to, nor recombination between, different species are well defined. As such do not allow inter-species mutation and have enforced $\rho = 1$.

Secondly, the performance as the performance of an individual agent is highly stochastic and dependent on the behaviour of other agents. In fact, our true fitness function is of the form $f(\mathbf{x}, \mathbf{P})$, where $\mathbf{P}$ is some population of agents, and is taken to be the average efficiency of the agent with parameters $\mathbf{x}$ over the course of a run amongst this population. Good agents in early (poorly optimised) populations are likely to have extremely high fitness as competition for mail is scarce. If we allow them to retain their fitness over many generations, this will strongly bias the ES towards their parameters. For this reason we evaluate the fitness of parents and children every generation in a population $\mathscr{P} \cup \mathscr{C}$.

3   Results

In this section, we discuss the numerical results. First we describe the general tendencies, presenting results representative of our update rules in the static and dynamic environments. We also test the rules' robustness to sudden change with the removal of all agents specialised in a particular mail type.

The second part of this section is dedicated to the optimisation of the parameters, and the selection of the best possible combination of update rules and threshold functions in terms of the overall efficiency.

3.1   General Tendencies

In this section, we investigate the performance of the different rules, both in terms of their behaviour under different conditions and in terms of absolute efficiency. However, investigating the influence of the system parameters on performance is beyond the scope of this paper. Therefore, we have fixed the parameters to a *standard setting*. In order to have a fair comparison between different environments we take and $R_{a/m} = 1$ in a static environment, while in the dynamic environment we take $R_{a/m} = 0.5$ (as $\overline{\pi_m(t)} = 0.5$ over a period) and we fix $N_m = 2$. The standard dynamic environment has a period $\xi = 50$. We simulate the system with $N_a = 5 \times 10^4$ agents as our investigations have shown this to be large enough to avoid finite size effects and to reduce the variance enough to remove the need for error bars.

We fix the various parameters to the following values: for the update rules $\theta_{min} = 0$, $\theta_{max} = 50$, $\varepsilon = \psi = 5$, $\alpha = \beta = \eta = 0.5$ and for the ETF $\lambda = 2$. A standard run consists of 500 iterations over which the average efficiency per agent is monitored. Note that all simulations are implemented in C++ and are performed on a Linux-PC cluster.

As a rule of thumb, we consider an agent to be fully specialised in a mail type if its threshold for this type is less than a distance of 1% of the possible range from θ_{min} while all other thresholds are within 1% of θ_{max}. The qualitative behaviour of the SO rule, as shown in figure 1 (top), is typical for all update rules although the speed of convergence and asymptotic values, depend on both the update rule and threshold function. We see that the algorithm accounts for the genesis of specialisation. The system tends towards a stable asymptotic regime in which most agents are specialised and the specialists are equally split between mail types. We see that $\ell.1$ is almost negligible while we have non-zero $\ell.2$ is indicative of the high value of θ_{max}. With the SO rule and $\theta_{min} = 0$, $\ell.2$ becomes impossible once an agent has

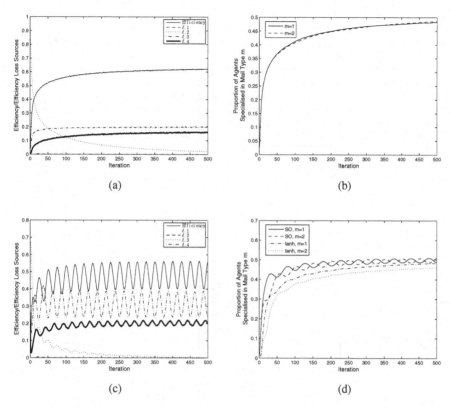

Fig. 1 (a): evolution of the efficiency and loss sources during a single run in the standard static environment using the SO rule. Note that $\ell.1$ is negligible everywhere and $\ell.2$ tends to 0, while $\ell.3$ and $\ell.4$ (and hence the efficiency) quickly tend to their long time values. (b): the population of agents tends towards an equal split in specialisation with almost all agents specialised. (c): evolution of the efficiency and loss sources during a single run in the standard dynamic environment using the SO rule. The values of the loss sources and the efficiency fluctuate around their *average* values, which are qualitatively similar to those in the static environment. (d): the difference in specialisation behaviour between the SO and the tanh rule. Note that the tanh rule tends to a static, uneven (initial condition dependent) set of specialisations, while the SO rule efficiently adapts to changes in the environment.

taken a piece of mail. Agents with all initial thresholds close to θ_{max} may never, over the course of a run, encounter a batch of mail with a strong enough stimulus to accept it.

In the dynamic environment (see figure 1, bottom), we observe that the efficiency fluctuates about some average value over the course of a wavelength. The qualitative behaviour, as shown in figure 1 for the SO rule, is typical for most update rules and the behaviour of their average values is qualitatively similar to that in the static environment. However, the specialisation that drives this behaviour varies between rules. In particular, all rules based on hidden variables ((m)tanh) tend to specialise

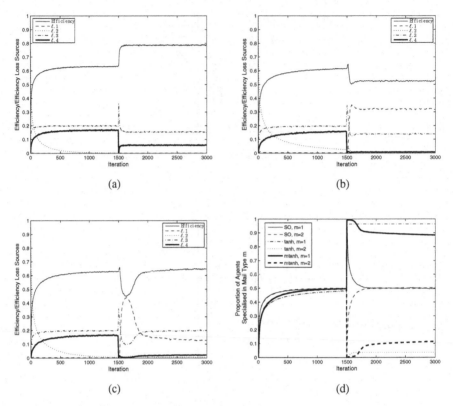

Fig. 2 Efficiency and loss sources in a static environment with removal of specialised agents. The SO rule (a) almost immediately returns to the optimal split in specialisations. Due to saturation, the tanh rule (b) is unable to re-specialise, resulting in a dramatic increase in consecutive changeovers ($\ell.1$). Although the mtanh rule (c) is capable of re-specialising, it does so far less efficiently than the SO rule and initially reacts similarly to the tanh rule. (d): evolution of the fraction of specialised agents for the various rules.

in a manner similar to the mtanh rule, while the other algorithms (VRT, SO and DH) behave qualitatively similarly to the SO rule.

The tanh function, like any sigmoid function, is effectively constant (i.e. saturated) for sufficiently large arguments. The saturation region is reached when an agent using the tanh update rule repeatedly takes the same mail type. Once in this region, the update rule becomes incapable of effective self reinforcement on which the VRT model relies, and incapable of reacting to changes in the environment. A similar problem can be encountered in neural networks with sigmoidal nodes, in which Hebbian learning drives synaptic weights into the saturation region of the function rendering the relative sizes of these weights meaningless [15] thus removing the selectivity of the node.

The tanh rule is the only rule inherently unable to dynamically adapt its thresholds due to the saturation effects described above, while the other rules can do so if given suitable parameters (i.e. a lowering of η in the mtanh rule). To highlight the effects of saturation for the tanh rule, we let the system equilibrate for 1500 iterations in the standard static environment to allow agents to specialise fully. Then we remove that half of the population that is most specialised in e.g. mail type 2, and equilibrate the remaining system (with halved $R_{a/m}$) for a further 1500 iterations. The results, shown in figure 2, show that in contrast to the SO rule, which adapts very quickly to the change, none of the specialised agents using the tanh rule re-specialise.

Stability is reached when the average stimulus of mail type 2 reaches a high enough level to force changeovers a significant proportion of the time, leading to high levels of $\ell.1$. The mtanh rule is able to somewhat adapt to this by lowering the thresholds of agents taking type 2 mail most often, causing some to re-specialise. Once enough agents are re-specialised that the average stimulus of type 2 mail drops to a level where frequent changeovers are unlikely the re-specialisation slows significantly meaning that an optimal set of specialisations will not be regained.

3.2   Evolutionary Optimisation

As explained in section 2.2, we employ an ES algorithm to obtain the optimal parameters and allow inter-update-rule competition. We use re-parametrisation to obtain better ES performance, both for parameters with fixed relationships with other ES variables ($p_1 = \theta_{max} - \theta_{min}, p_2 = \alpha/p_1, p_3 = \beta/p_1$) and for those with exponential dependence ($p_4 = \log(\lambda)$). Parameters are constrained to $\theta_{min}, p_1, \varepsilon, \psi \in [0, 100]$, $p_4 \in [\log(1), \log(10)]$, and $\eta \in [-1, 1]$. It turns out that the optimised SO rule outperforms the other update rules in virtually all circumstances. The only update rules that can compete with it are those that can effectively mimic its behaviour by extreme choices of parameters. As this against the spirit of the nature inspired VRT rule, we have constrained its parameters to $p_2, p_3 \in [0, 0.5]$.

In the ES we use $\mu, \ell = 5000$ and initialise each element of σ to the size of parameter space. As mtanh can evolve into tanh for $\eta = 1$, we do not explicitly use the tanh rule in our algorithm. The ES was run for 100 generations and results are averaged over 50 runs. We investigate the average fitness both of all the agents competing in an ES generation, but also of the parent agents (i.e. those selected by the ES as parents for the next generation). We also look at the relative fractions of each update rule within the population.

In the both environments (see figure 3), the ES quickly obtains a high level of efficiency. This is obtained by dropping θ_{max} to a much lower value than intuitively expected (and used in the standard setting). The remaining efficiency gain is mainly a consequence of the increasing fraction of the population with a good rule set (see figure 4). We observe a peak in the efficiency of parent agents during the early generations as agents first discover good rules and parameter sets. These well optimised

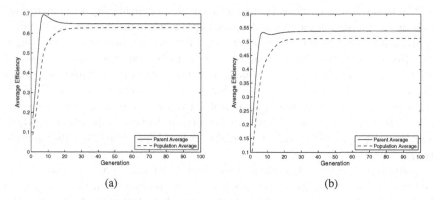

Fig. 3 Evolution of the efficiency during an ES optimisation of the population of agents in the static (a) and dynamic (b) environment. In environments both ES leads to increase the efficiency of both on average and of parent agents. After ≈ 30 generations efficiencies tend to a stable value. We also observe a peak in parent efficiency early in the run.

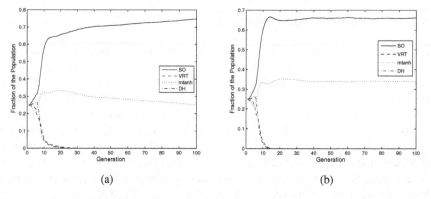

Fig. 4 Evolution of the efficiency during an ES optimisation of the population of agents in the static (a) and dynamic (b) environment. In both environments, all update rules eventually tend to extinction, except the mtanh and SO rule which effectively become the same. The relative fractions are determined by the initial conditions and sensitivity to mutations.

agents have more mail available than those in later generations, as they are competing with inferior agents.

We see from figure 4 that the two best update rules are the SO and the mtanh rule (with $\eta < 0$ and large ε, ψ, thus approximating SO). The SO rule, however, has the added advantage of not being able to move away from this behaviour. As such, in both environments, the SO rule has an initial advantage over mtanh. In the static environment the ability of mtanh to mutate away from SO style behaviour causes the fraction the mtanh agents to be slow out-competed by SO, although in the dynamic environment we see no bias towards either rule.

Table 1 The average efficiency of the different methods

Environment	Static	Dynamic
VRT	0.501	0.412
SO	0.586	0.467
ES	0.629	0.512

An analysis of the inter-run variance shows that the final proportions of SO and mtanh rules are only an average trends, and in some runs the fraction of mtanh agents increases from its early value, while in others it decreases. The variance is low for the first ≈ 10 generations, while VRT and DH become extinct and mtanh finds SO-like behaviour, but subsequently increases. In fact the final proportions ± 1 standard deviation are 0.747 ± 0.257 (0.661 ± 0.189) for the SO rule and 0.253 ± 0.257 (0.339 ± 0.189) for the mtanh rule in the static (dynamic) environment.

In Table 1, we illustrate the effect on the efficiency made by the introduction of new update rules and evolutionary optimisation in comparison to the original VRT model, which already outperforms a range of other general purpose algorithms [12, 13]. These efficiencies are given in the static and dynamic environment. Efficiencies are averaged over 500 iterations (including the initial specialisation period). ES results are given as an average over 50 runs of the performance of all the agents in the final (100th) generation. Note that these figures include the damage caused by mutations and the introduction of random agents. The SO rule already provides a large improvement over the VRT rule, while the ES determined rules and parameters obtain increased performance, particularly in the dynamic environment.

4   Conclusions and Outlook

In this paper, we have studied an agent based model for distributed mail retrieval. The efficiency and flexibility have been investigated both in static and dynamic environments. and with respect to catastrophic breakdowns of agents. We have introduced new rules for mail selection and specialisation and have used a evolutionary algorithm to optimise these further. We have shown that some of the new rules have improved performance compared to existing ones. The best ones give increased efficiency by 25.5% in a static, and 24.3% in a dynamic environment, compared to a method (VRT) which already outperformed a variety of other algorithms [12].

Although speciation and extinction do occur in the current model using evolution strategies, proper self-organising behaviour such as cooperation between the agents is not observed. The main limitation of the current model is the random choice of cities which does not really allow agents to develop cooperative strategies, and direct competition is the only driving force behind the evolution of species. Secondly, the dynamics of the interactions in the ES may still be limited by our choice of the functional forms of the new rules. Hence it would be interesting to apply genetic programming in which agents are allowed to develop their strategies freely.

References

1. Beyer, H., Schwefel, H.: Evolution strategies - a comprehensive introduction. Natural Computing 1(1), 3–52 (2002)
2. Bonabeau, E., Dorigo, M., Theraulaz, G.: Swarm Intelligence: From Natural to Artificial Systems. Oxford University Press, Oxford (1999)
3. Bonabeau, E., Guérin, S., Snyers, D., Kuntz, P., Theraulaz, G.: Three-dimensional architectures grown by simple stigmergic agents. Biosystems 56(1), 13–32 (2000)
4. Bonabeau, E., Sobkowski, A., Theraulaz, G., Deneubourg, J.-L.: Adaptive task allocation inspired by a model of division of labor in social insects. In: Biocomputing and Emergent Computation, pp. 36–45 (1997)
5. Bonabeau, E., Theraulaz, G., Deneubourg, J.-L.: Fixed response thresholds and the regulation of division of labour in insect societies. Bulletin of Mathematical Biology 60, 753–807 (1998)
6. Campos, M., Bonabeau, E., Thraulaz, G., Deneubourg, J.-L.: Dynamic scheduling and division of labor in social insects. Adaptive Behavior 8(2), 83–92 (2001)
7. Chevaleyre, Y., Dunne, P.E., Endriss, U., Lang, J., Lemaître, M., Maudet, N., Padget, J., Phelps, S., Rodríguez-Aguilar, J.A., Sousa, P.: Issues in multiagent resource allocation
8. Goldingay, H., van Mourik, J.: The influence of memory in a threshold model for distributed task assignment. In: SASO 2008: Proceedings of the 2008 Second IEEE International Conference on Self-Adaptive and Self-Organizing Systems, Washington, DC, USA, pp. 117–126. IEEE Computer Society, Los Alamitos (2008)
9. Grassé, P.P.: La reconstruction du nid et les interactions inter-individuelles chez les bellicositermes natalenis et cubitermes sp. la théorie de la stigmergie: essai d'interprétation des termites constructeurs. Insectes Sociaux 6, 41–83 (1959)
10. Hong, B., Prasanna, V.K.: Distributed adaptive task allocation in heterogeneous computing environments to maximize throughput. In: IPDPS (2004)
11. Low, K.H., Leow, W.K., Ang Jr., M.H.: Task allocation via self-organizing swarm coalitions in distributed mobile sensor network. In: Proc. 19th National Conference on Artificial Intelligence (AAAI 2004), pp. 28–33 (2004)
12. Price, R.: Evaluaton of adaptive nature inspired task allocation against alternate decentralised multiagent strategies (2004)
13. Price, R., Tiňo, P.: Evaluation of Adaptive Nature Inspired Task Allocation Against Alternate Decentralised Multiagent Strategies. In: Yao, X., Burke, E.K., Lozano, J.A., Smith, J., Merelo-Guervós, J.J., Bullinaria, J.A., Rowe, J.E., Tiňo, P., Kabán, A., Schwefel, H.-P. (eds.) PPSN 2004. LNCS, vol. 3242, pp. 982–990. Springer, Heidelberg (2004)
14. Rana, O.F., Stout, K.: What is scalability in multi-agent systems? In: AGENTS 2000: Proceedings of the fourth international conference on Autonomous agents, pp. 56–63 (2000)
15. Schraudolf, N.N., Sejnowski, T.J.: Unsupervised discrimination of clustered data via optimization of binary information gain. Advances in Neural Information Processing Systems 5, 499–506 (1993)
16. Theraulaz, G., Bonabeau, E., Deneubourg, J.-L.: Response threshold reinforcement and division of labour in insect societies. Proceedings of the Royal Society B: Biological Sciences 265, 327–332 (1998)

Efficient Mining of High Utility Patterns over Data Streams with a Sliding Window Method

Chowdhury Farhan Ahmed, Syed Khairuzzaman Tanbeer, and Byeong-Soo Jeong

Abstract. High utility pattern (HUP) mining over data streams has become a challenging research issue in data mining. The existing sliding window-based HUP mining algorithms over stream data suffer from the level-wise candidate generation-and-test problem. Therefore, they need a large amount of execution time and memory. Moreover, their data structures are not suitable for interactive mining. To solve these problems of the existing algorithms, in this paper, we propose a new tree structure, called HUS-tree (High Utility Stream tree) and a novel algorithm, called HUPMS (HUP Mining over Stream data), for sliding window-based HUP mining over data streams. By capturing the important information of the stream data into an HUS-tree, our HUPMS algorithm can mine all the HUPs in the current window with a pattern growth approach. Moreover, HUS-tree is very efficient for interactive mining. Extensive performance analyses show that our algorithm significantly outperforms the existing sliding window-based HUP mining algorithms.

1 Introduction

Traditional frequent pattern mining algorithms [1], [7], [6], [9], [16] only consider the binary (0/1) frequency values of items in transactions and same profit value for every item. However, the customer may purchase more than one of the same item, and the unit price may vary among items. High utility pattern mining approaches [19], [18], [14], [4], [12] have been proposed to overcome this problem. By considering non-binary frequency values of items in transactions and different profit values of every item, HUP mining becomes a very important research issue in data mining and knowledge discovery. With HUP mining, several important business area decisions can be made to maximize revenue, minimize marketing or reduce inventory. Moreover, customers/itemsets contributing the most profit can be identified. In addition to real world retail markets, we can also consider biological gene databases

Chowdhury Farhan Ahmed, Syed Khairuzzaman Tanbeer, and Byeong-Soo Jeong
Database Lab, Department of Computer Engineering, Kyung Hee University
1 Seochun-dong, Kihung-gu, Youngin-is, Kyunggi-do, 446-701, Republic of Korea
e-mail: farhan,tanbeer,jeong@khu.ac.kr

Roger Lee (Ed.): SNPD 2010, SCI 295, pp. 99–113, 2010.
springerlink.com © Springer-Verlag Berlin Heidelberg 2010

and web click streams, where the importance of each gene or website is different and their occurrences are not limited to a 0/1 value.

On the other hand, in recent years, many applications generate data streams in real time, such as sensor data generated from sensor networks, transaction flows in retail chains, web click streams in web applications, performance measurement in network monitoring and traffic management, call records in telecommunications, and so on. A data stream is a continuous, unbounded and ordered sequence of items that arrive in order of time. Due to this reason, it is impossible to maintain all the elements of a data stream. Moreover, in a data stream, old information may be unimportant or obsolete in the current time period. Accordingly, it is important to differentiate recently generated information from the old information. Sliding window-based methods [8], [2], [11], [13] have been proposed to extract the recent change of knowledge in a data stream adaptively in the context of traditional frequent pattern mining.

However, the existing sliding window-based HUP mining algorithms [17], [3], [10] are not efficient. They suffer from the level-wise candidate generation-and-test problem. Therefore, they generate a large number of candidate patterns. Recently, two algorithms, MHUI-BIT and MHUI-TID [10], have been proposed for sliding window-based HUP mining. The authors showed their algorithms have outperformed all the previous algorithms. However, these algorithms also suffer from the level-wise candidate generation-and-test problem. To reduce the calculation time, they have used bit vectors and TID-lists for each distinct item. But these lists become very large and inefficient when the numbers of distinct items and/or transactions become large in a window. They have also used a tree structure to store only *length-1* and *length-2* candidate patterns. Hence, candidate patterns having length greater than two must be calculated using an explicit level-wise candidate generation method.

Moreover, the data structures of the existing sliding-window based HUP mining algorithms do not have the "*build once mine many*" property (by building the data structure only once, several mining operations can be done) for interactive mining. As a consequence, they cannot use their previous data structures and mining results for the new mining threshold. In our real world, however, the users need to repeatedly change the minimum threshold for useful information extraction according to their application requirements. Therefore, the "*build once mine many*" property is essentially needed to solve these interactive mining problems.

Motivated by these real world scenarios, in this paper, we propose a new tree structure, called HUS-tree (High Utility Stream tree) and a novel algorithm, called HUPMS (HUP Mining over Stream data), for sliding window-based HUP mining over data streams. Our HUPMS algorithm can capture important information from a data stream in a batch-by-batch fashion inside the nodes of an HUS-tree. Due to this capability of an HUS-tree, HUPMS can save a significant amount of processing time for removing the old batch information when a window slides. By exploiting a pattern growth approach, HUPMS can successfully mine all the resultant patterns. Therefore, it can avoid the level-wise candidate generation-and-test problem completely and reduces a large number of candidate patterns. As a consequence, it

ITEM TID	a	b	c	d	e	Trans. Utility($)
T_1	0	0	0	3	4	64
T_2	2	0	8	0	0	28
T_3	0	2	8	2	0	52
T_4	4	8	0	0	0	56
T_5	0	3	0	0	2	38
T_6	6	5	0	4	0	74

ITEM	PROFIT($) (per unit)
a	2
b	6
c	3
d	8
e	10

(a) Transaction Database   (b) Utility Table

Fig. 1 Example of a transaction database and utility table

significantly reduces the execution time and memory usage for stream data processing. Moreover, an HUS-tree has the "*build once mine many*" property for interactive
mining. Extensive performance analyses show that our algorithm outperforms the
existing algorithms, efficient for interactive mining, can efficiently handle a large
number of distinct items and transactions.

The remainder of this paper is organized as follows. In Section 2, we describe the
background and related work. In Section 3, we develop our proposed tree structure
and algorithm. In Section 4, our experimental results are presented and analyzed.
Finally, in Section 5, conclusions are drawn.

2 Background and Related Work

We adopted definitions similar to those presented in the previous works [19], [18],
[14]. Let $I = \{i_1, i_2, \ldots\ldots i_m\}$ be a set of items and D be a transaction database
$\{T_1, T_2, \ldots\ldots T_n\}$ where each transaction $T_i \in D$ is a subset of I.

Definition 1. The internal utility or local transaction utility value $l(i_p, T_q)$, represents the quantity of item i_p in transaction T_q. For example, in Fig. 1(a), $l(c, T_2) = 8$.

Definition 2. The external utility $p(i_p)$ is the unit profit value of item i_p. For example, in Fig. 1(b), $p(c) = 3$.

Definition 3. Utility $u(i_p, T_q)$, is the quantitative measure of utility for item i_p in
transaction T_q, defined by

$$u(i_p, T_q) = l(i_p, T_q) \times p(i_p) \tag{1}$$

For example, $u(c, T_2) = 8 \times 3 = 24$ in Fig. 1.

Definition 4. The utility of an itemset X in transaction T_q, $u(X, T_q)$ is defined by,

$$u(X, T_q) = \sum_{i_p \in X} u(i_p, T_q) \tag{2}$$

where $X = \{i_1, i_2, \ldots\ldots i_k\}$ is a k-itemset, $X \subseteq T_q$ and $1 \leq k \leq m$. For example,
$u(ab, T_6) = 6 \times 2 + 5 \times 6 = 42$ in Fig. 1.

Definition 5. The utility of an itemset X is defined by,

$$u(X) = \sum_{T_q \in D} \sum_{i_p \in X} u(i_p, T_q) \tag{3}$$

For example, $u(ab) = u(ab, T_4) + u(ab, T_6) = 56 + 42 = 98$ in Fig. 1.

Definition 6. The transaction utility of transaction T_q denoted as $tu(T_q)$ describes the total profit of that transaction and it is defined by,

$$tu(T_q) = \sum_{i_p \in T_q} u(i_p, T_q) \tag{4}$$

For example, $tu(T_1) = u(d, T_1) + u(e, T_1) = 24 + 40 = 64$ in Fig. 1.

Definition 7. The minimum utility threshold δ, is given by the percentage of total transaction utility values of the database. In Fig. 1, the summation of all the transaction utility values is 312. If δ is 30% or we can also express it as 0.3, then the minimum utility value can be defined as

$$minutil = \delta \times \sum_{T_q \in D} tu(T_q) \tag{5}$$

Therefore, in this example $minutil = 0.3 \times 312 = 93.6$ in Fig. 1.

Definition 8. An itemset X is a high utility itemset, if $u(X) \geq minutil$. Finding high utility itemsets means determining all itemsets X having criteria $u(X) \geq minutil$.

For frequent pattern mining, the *downward closure* property [1] is used to prune the infrequent patterns. This property says that if a pattern is infrequent, then all of its super patterns must be infrequent. However, the main challenge of facing high utility pattern mining areas is the itemset utility does not have the *downward closure* property. For example, if $minutil = 93.6$ in Fig. 1, then "a" is a low utility item as $u(a) = 24$. However, its superpattern "ab" is a high utility itemset as $u(ab) = 98$. As a consequence, this property is not satisfied here. We can maintain the *downward closure* property by transaction weighted utilization.

Definition 9. The transaction weighted utilization of an itemset X, denoted by $twu(X)$, is the sum of the transaction utilities of all transactions containing X.

$$twu(X) = \sum_{X \subseteq T_q \in D} tu(T_q) \tag{6}$$

For example, $twu(c) = tu(T_2) + tu(T_3) = 28 + 52 = 80$ in Fig. 1. The *downward closure* property can be maintained using transaction weighted utilization. Here, for $minutil = 93.6$ in Fig. 1 as $twu(c) < minutil$, any super pattern of "c" cannot be a high twu itemset (candidate itemset) and obviously cannot be a high utility itemset.

Definition 10. X is a high transaction weighted utilization itemset (i.e., a candidate itemset) if $twu(X) \geq minutil$.

Now we describe the related work of HUP mining. The theoretical model and definitions of HUP mining were given in [19]. This approach is called as mining with expected utility (MEU). Later, the same authors proposed two new algorithms, UMining and UMining_H, to calculate high utility patterns [18]. However, these methods do not satisfy the *downward closure* property of Apriori [1] and overestimate too many patterns. The Two-Phase [14] algorithm was developed based on the definitions of [19] for HUP mining using the *downward closure* property with the measure of transaction weighted utilization (*twu*). The isolated items discarding strategy (IIDS) [12] for discovering high utility patterns was proposed to reduce some candidates in every pass of databases. Applying IIDS, the authors developed two efficient HUP mining algorithms called FUM and DCG+. But, these algorithms suffer from the problem of level-wise candidate generation-and-test methodology and need several database scans.

Several sliding window-based stream data mining algorithms [8], [2], [11], [13] have been developed in the context of traditional frequent pattern mining. Tseng et al. [17], [3] proposed the first sliding window-based HUP mining algorithm, called THUI-Mine (Temporal High Utility Itemsets Mine). It is based on the Two-Phase algorithm and therefore suffers from the level-wise candidate generations and test methodology. Recently, two algorithms, called MHUI-BIT (Mining High Utility Itemsets based on BITvector) and MHUI-TID (Mining High Utility Itemsets based on TIDlist) [10], have been proposed for sliding window-based HUP mining which are better than the THUI-Mine algorithm. It is shown in [10] that the MHUI-TID algorithm is always better than the other algorithms. However, limitations of these existing algorithms have been discussed in Section 1. In this paper, we proposed a novel tree-based algorithm to address the limitations of the existing algorithms.

3 HUPMS: Our Proposed Algorithm

3.1 Preliminaries

Transaction stream is a kind of data stream which occurs in many application areas such as retail chain, web click analysis, etc. It may have infinite number of transactions. A batch of transactions contains a nonempty set of transactions. Figure 2 shows an example of transaction stream divided into four batches with equal length (it uses the utility table of Fig. 1(b)). A window consists of multiple batches. In our example, we assume that one window contains three batches of transactions (we denote the *i*-th window as W_i and the *j*-th batch as B_j). That is, W_1 contains B_1, B_2 and B_3. Similarly W_2 contains B_2, B_3 and B_4.

Definition 11. The utility of an itemset X in a batch B_j is defined by,

$$u_{B_j}(X) = \sum_{T_q \in B_j} \sum_{i_p \in X \subseteq T_q} u(i_p, T_q) \tag{7}$$

For example, $u_{B_4}(ab) = u(ab, T_7) + u(ab, T_8) = 16 + 32 = 48$ in Fig. 2.

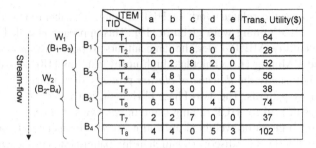

Fig. 2 Example of a transaction data stream

Definition 12. The utility of an itemset X in a window W_k is defined by,

$$u_{W_k}(X) = \sum_{B_j \in W_k} \sum_{X \subseteq T_q \in B_j} u(X, B_j) \tag{8}$$

For example, $u_{W_2}(ab) = u_{B_2}(ab) + u_{B_3}(ab) + u_{B_4}(ab) = 56 + 42 + 48 = 146$ in Fig. 2.

Definition 13. The minimum utility threshold δ_{W_k} is given by percentage of the total transaction utility values of window W_k. In Fig. 2, the summation of all the transaction utility values in W_2 is 359. If δ_{W_k} is 30%, then minimum utility value in this window can be defined as

$$minutil_{W_k} = \delta_{W_k} \times \sum_{T_q \in W_k} tu(T_q) \tag{9}$$

So, in this example $minutil_{W_2} = 0.3 \times 359 = 107.7$ in Fig. 2.

Definition 14. A pattern X is a high utility pattern in window W_k, if $u_{W_k}(X) \geq minutil_{W_k}$. Finding high utility patterns in window W_k means find out all the patterns X having criteria $u_{W_k}(X) \geq minutil_{W_k}$. For $minutil_{W_2} = 107.7$ pattern "ab" is a high utility pattern, as $u_{W_2}(ab) = 146$.

Definition 15. The transaction weighted utilization of an itemset X in a window W_k, denoted by $twu_{W_k}(X)$, is similarly defined as utility of a pattern X in W_k. For example, $twu_{W_2}(c) = twu_{B_2}(c) + twu_{B_4}(c) = 52 + 37 = 89$ in Fig. 2. X is a high transaction weighted utilization itemset in W_k if $twu_{W_k}(X) \geq minutil_{W_k}$. However, the *downward-closure* property can also be maintained using transaction weighted utilization in a window. Here, for $minutil_{W_2} = 107.7$ in Fig. 2 as $twu_{W_2}(c) < minutil_{W_2}$, any super pattern of item "c" cannot be a candidate pattern and obviously can not be a HUP in this window.

3.2 HUS-Tree Construction

In this section, we describe the construction process of our proposed tree structure, HUS-tree (High Utility Stream tree) to capture stream data. It arranges the items in

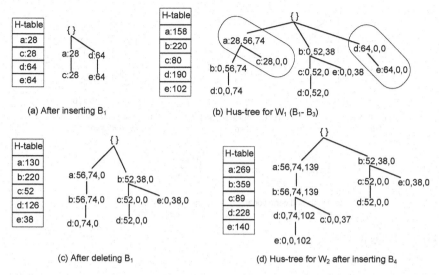

Fig. 3 HUS-tree construction for W_1 and W_2

lexicographic order. A header table is maintained to keep an item order in our tree structure. Each entry in a header table explicitly maintains *item-id* and *twu* value of an item. However, each node in a tree maintains *item-id* and batch-by-batch *twu* information to efficiently maintain the window sliding environment. To facilitate the tree traversals, adjacent links are also maintained (not shown in the figures for simplicity) in our tree structure.

Consider the example data stream of Fig. 2. We scan the transactions one by one, sort the items in a transaction according to header table item order (*lexicographic order*) and then insert into the tree. The first transaction T_1 has a *tu* value of 64 and contains two items "*d*" and "*e*". At first item "*d*" is inserted into HUS-tree by creating a node with *item-id* "*d*" and *twu* value 64. Subsequently, item "*e*" is inserted with *twu* value 64 as its child. This information is also updated in the header table. After inserting T_1 and T_2 into HUS-tree. Figure 3(a) shows the HUS-tree constructed for B_1 (T_1 and T_2). Similarly, B_2 and B_3 are inserted as shown in Fig. 3(b). Since W_1 contains the first three batches, Fig 3(b) is the final tree for it. However, these figures clearly indicate that each node in the tree contains batch-by-batch *twu* information. For example, Figure 3(b) indicates that in path "*d e*", both items "*d*" and "*e*" only occur in B_1 but not in B_2 and B_3.

When the data stream moves to B_4, it is necessary to delete the information of B_1, because B_1 does not belong to W_2 and therefore the information of B_1 becomes garbage in W_2. In order to delete the information of B_1 from the tree, the *twu* counters of the nodes are shifted one position left to remove the *twu* information of B_1 and include the *twu* information for B_4. Figure 3(b) also indicates (by rounding rectangles) the two transactions of B_1 in the tree. Since node with *item-id* "*a*" contains information for B_2 and B_3, its new information is now $\{a : 56, 74, 0\}$. On the other hand, its child "*c*" does not contain any information for B_2 and B_3 and therefore

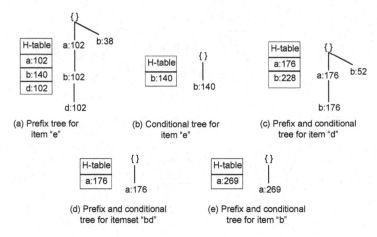

Fig. 4 Mining process

it becomes $\{c : 0,0,0\}$ after the shifting operation. As a consequence, it is deleted from the tree. Similarly, path "*d e*" is also deleted from the tree. Figure 3(c) shows the tree after deleting B_1. Subsequently, B_4 is inserted into the tree. Hence, now the three *twu* information of each node represents B_2, B_3 and B_4. Figure 3(d) shows the tree after inserting B_4. As W_2 contains B_2, B_3 and B_4, Fig 3(d) is the final tree for it.

3.3 Mining Process

In this section, we describe the mining process of our proposed HUPMS algorithm. To apply a pattern growth mining approach, HUPMS first creates a prefix tree for the bottom-most item. To create it, all the branches prefixing that item are taken with the *twu* of that item. For the mining purpose, we add all the *twu* values of a node in the prefix tree to indicate its total *twu* value in this current window. Subsequently, conditional tree for that item is created from the prefix tree by eliminating the nodes containing items having low *twu* value with that particular item.

Suppose we want to mine the recent HUPs in the data stream presented at Fig. 2. It means we have to find out all the HUPs in W_2. Consider $\delta_{W_2} = 30\%$ and $minutil_{W_2} = 107.7$ accordingly. We start from the bottom-most item "*e*". The prefix tree of item "*e*" is shown in Fig. 4(a). It shows that, items "*a*" and "*d*" cannot form any candidate patterns with item "*e*" as they have low *twu* value with "*e*". Hence, the conditional tree of item "*e*" is derived by deleting all the nodes containing items "*a*" and "*d*" from the prefix tree of "*e*" as shown in Fig. 4(b). Candidate patterns (1) $\{b,e : 140\}$, and (2) $\{e : 140\}$ are generated here. Prefix tree of item "*d*" is created in Fig. 4(c). Both items "*a*" and "*b*" have high *twu* values with "*d*" and therefore this tree is also the conditional tree for item "*d*". However, candidate patterns (3) $\{a,d : 176\}$, (4) $\{b,d : 228\}$ and (5) $\{d : 228\}$ are generated. Subsequently, prefix and conditional tree for itemset "*bd*" is created in Fig. 4(d) and candidate patterns (6) $\{a,b,d : 176\}$ is generated.

The *twu* value of item "*c*" (89) is less than $minutil_{W_2}$. According to the *downward closure* property any super pattern of it cannot be a candidate pattern as well as a HUP. Therefore, we do not need to create any prefix/conditional tree for item "*c*". Prefix and conditional tree for item "*b*" is created in Fig. 4(e) and candidate patterns (7) $\{a,b : 269\}$ and (8) $\{b : 359\}$ are generated. The last candidate pattern (9) $\{a : 269\}$ is generated for the top-most item "*a*". A second scan of W_2 is required to find high utility patterns from these candidate patterns. For this example, the actual high utility patterns with their actual utility value are as follows $\{b,d : 154\}$, $\{a,b,d : 146\}$, $\{a,b : 134\}$ and $\{b : 144\}$. Finally, pseudo-code of the HUPMS algorithm is presented in Fig. 5.

3.4   Interactive Mining

HUS-tree has the "*build once mine many*" property for interactive mining. For example, after the creation of HUS-tree in Fig. 3(d) for W_2, at first we can mine the resultant patterns for $\delta_{W_2} = 30\%$. Subsequently, we can again mine the resultant patterns for different minimum thresholds (like 20%, 25%, 40% etc.) without rebuilding the tree. Hence, after the first time, we do not have to create the tree. Advantages in mining time and second scanning time of the current window for determining high utility patterns from high *twu* patterns are also realized. For example, after obtaining four high utility patterns from nine candidate patterns for $\delta_{W_2} = 30\%$, if we perform mining operation for larger values of δ_{W_2}, then the candidate set is a subset of the previous candidate set and, the actual high utility patterns can be defined without mining or scanning the current window again. If δ_{W_2} is smaller than the previous trial, then the candidate set is a super set of the previous candidate set. Therefore, after mining, the current window can be scanned only for patterns that did not appear in the previous candidate set.

4   Experimental Results

To evaluate the performance of our proposed algorithm, we have performed several experiments on IBM synthetic dataset *T10I4D100K* and real life datasets *BMS-POS*, *kosarak* from frequent itemset mining dataset repository [5]. These datasets do not provide profit values or the quantity of each item for each transaction. As for the performance evaluation of the previous utility based pattern mining [14], [4], [12], [17], [3], [10], we generated random numbers for the profit values of each item and quantity of each item in each transaction, ranging from 1.0 to 10.0 and 1 to 10, respectively. Based on our observation in real world databases that most items carry low profit, we generated the profit values using a lognormal distribution. Some other high utility pattern mining research [14], [4], [12], [17], [3], [10] has adopted the same technique. Finally, we present our result by using a real-life dataset (*Chainstore*) with real utility values [15]. Since the existing MHUI-TID algorithm outperforms the other sliding window-based HUP mining algorithms, we compare the performance of our algorithm with only the MHUI-TID algorithm. Our programs

Input: A transaction data stream, utility table, δ_{W_i}, batch size, window size.
Output: High utility patterns for the current window.

begin

    Create the global header table H to keep the items in the *lexicographic* order

    foreach *batch B_j* **do**

        Delete obsolete batches for current window W_i from the HUS-tree if required

        foreach *transaction T_k in batch B_j* **do**

            Sort the items of T_k in the order of H

            Update *twu* in the header table H

            Insert T_k in the HUS-tree

        end

        while *any mining request from the user* **do**

            Input δ_{W_i} from the user

            foreach *each item α from the bottom of H* **do**

                Add pattern α in the candidate pattern list

                Create Prefix tree PT_α with its header table HT_α for item α

                Call Mining $(PT_\alpha, HT_\alpha, \alpha)$

            end

            Calculate the HUPs from the candidate list

        end

    end

end

Procedure Mining(T, H, α)

begin

    foreach *item β of H* **do**

        if *twu(β)* < *minutil* **then**

            Delete β from T and H to create conditional tree CT and its header table HC

        end

    end

    foreach *item β in HC* **do**

        Add pattern $\alpha\beta$ in the candidate pattern list

        Create Prefix-tree $PT_{\alpha\beta}$ and Header table $HT_{\alpha\beta}$ for pattern $\alpha\beta$

        Call Mining $(PT_{\alpha\beta}, HT_{\alpha\beta}, \alpha\beta)$

    end

end

Fig. 5 The HUPMS algorithm

were written in Microsoft Visual C++ 6.0 and run with the Windows XP operating system on a Pentium dual core 2.13 GHz CPU with 2 GB main memory.

4.1 Runtime Efficiency of the HUPMS Algorithm

In this section, we show the runtime efficiency of HUPMS algorithm over the existing MHUI-TID algorithm using the *T10I4D100K* and *BMS-POS* datasets. Our

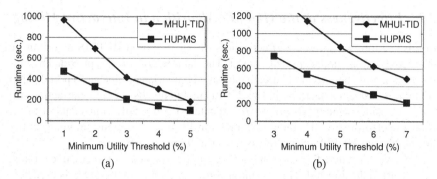

Fig. 6 Runtime comparison on the (a) *T10I4D100K* dataset (b) *BMS-POS* dataset

HUPMS algorithm keeps separate information for each batch inside the HUS-tree nodes. Hence, when the current window slides to a new window it can easily discard the information of obsolete batches from the tree nodes to update the tree. After that, it inserts the information of the new batches into the tree. Therefore, it is very efficient in sliding window-based stream data processing. Moreover, it exploits a pattern growth mining approach for mining the resultant HUPs in a window and significantly reduces the number of candidate patterns. Due to these reasons, it significantly outperforms the existing MHUI-TID algorithm in sliding window-based stream data mining with respect to execution time.

The *T10I4D100K* dataset was developed by the IBM Almaden Quest research group and obtained from the frequent itemset mining dataset repository [5]. This dataset contains 100,000 transactions and 870 distinct items. Its average transaction size is 10.1. Figure 6(a) shows the effect of sliding window-based stream data mining in this dataset. Here window size $W = 3B$ and batch size $B = 10K$, i.e., each window contains 3 batches and each batch contains 10K transactions. Accordingly, W_1 contains the first three batches (the first 30K transactions). A mining operation is done in this window using $\delta_{W_1} = 5\%$. Subsequently, the window slides to the next one and W_2 contains batches from B_2 to B_4. A mining operation is done again in this window using the same minimum threshold ($\delta_{W_2} = 5\%$). Other windows are processed similarly. As the minimum utility threshold is same in all the windows, we can refer it as δ. However, different performance curves can be found by varying the value of δ. The *x-axis* of Fig. 6(a) shows different values of δ and the *y-axis* shows the runtime. Figure 6(a) shows that our algorithm significantly outperforms the existing algorithm. Moreover, it also demonstrates that the runtime difference increases when the minimum utility threshold decreases.

Subsequently, we compare the performance of our algorithm on the *BMS-POS* dataset [20], [5]. It contains 515,597 transactions and 1,657 distinct items. Its average transaction size is 6.53. Figure 6(b) shows the effect of sliding window-based stream data mining in this dataset. Here window size $W = 3B$ and batch size $B = 30K$ are used. Figure 6(b) also shows that our algorithm significantly outperforms the existing algorithm.

4.2 Effectiveness of the HUPMS Algorithm in Interactive Mining

In this section, we show the effectiveness of our proposed HUPMS algorithm in interactive mining in the *kosarak* dataset. The dataset *kosarak* was provided by Ferenc Bodon and contains click-stream data of a Hungarian on-line news portal [5]. It contains 990,002 transactions and 41,270 distinct items. Its average transaction size is 8.1. Consider window size $W = 5B$ and batch size $B = 50$K in this dataset. If we want to perform several mining operations in W_1 using different minimum utility thresholds, our HUPMS can show its power of interactive mining.

We have taken here the result in Fig. 7(a) for the worst case of interactive mining, i.e., we listed the thresholds in descending order. A δ range of 4% to 8% is used here and we use $\delta = 8\%$ at first, then 7% and so on. As discussed in Section 3.4, we do not have to rebuild our HUS-tree after the first threshold, i.e., once the tree is created several mining operations can be performed without rebuilding it. As the threshold decreases, mining operation and second scan of the current window are needed. Therefore, we need to contruct the tree only for the first mining threshold, i.e., 8%. As a consequence, tree construction time is added with mining and second scanning time of the current window for $\delta = 8\%$. For the other thresholds, only mining and second scanning time of the current window are counted.

However, as candidates of threshold 7% are a superset of the candidates of threshold 8% , we can easily save the second scanning time of the current window for the candidates of threshold 8%. High utility itemsets from them are already calculated for threshold 8%. We have to scan the current window a second time only for the candidates newly added for threshold 7%. The best case occurs if we arrange the mining threshold in increasing order. Then candidates of threshold 6% are a subset of the candidates of threshold 5%. Therefore, without mining and a second scan of the current window, we can find the resultant high utility itemsets from the previous result. In that case, after the first mining threshold, the computation time for other thresholds is negligible compared to that required for the first. Figure 7(a) shows that the overall runtime of the existing algorithm becomes huge due to the large number of distinct items in the *kosarak* dataset. This figure also shows that our algorithm is efficient in interactive mining and significantly outperforms the existing algorithm.

4.3 Effect of Window Size Variation

For a sliding window-based stream data mining algorithm, runtime and memory requirement are dependent on the window size. A window consists of M batches and a batch consists of N transactions. Therefore, window size may vary depending on the number of batches in a window and the number of transactions in a batch. In this section, we analyze the performance of HUPMS by varying both of these two parameters over a real-life dataset with real utility values.

The real-life dataset was adopted from NU-MineBench 2.0, a powerful benchmark suite consisting of multiple data mining applications and databases [15]. The dataset called *Chain-store* was taken from a major chain in California and contains 1,112,949 transactions and 46,086 distinct items [15]. Its average transaction size

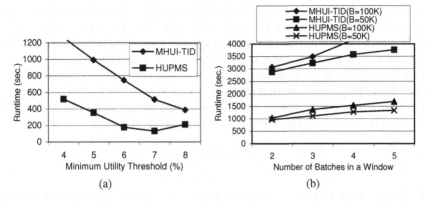

Fig. 7 (a) Effectiveness of interactive mining on the *kosarak* dataset (b) Effect of window size variation on the *Chain-store* dataset

is 7.2. We compare the performance of our algorithm with the existing MHUI-TID using this dataset with different window sizes changing the number of batches in a window and the number of transactions in a batch (Fig. 7(b)). The *x-axis* of Fig. 7(b) shows different window sizes containing 2, 3, 4 and 5 batches. Two curves of HUPMS algorithm and two curves for MHUI-TID algorithm show the effects of two different sizes of batches (50K and 100K). The mining operation is performed at each window with a minimum threshold (δ) value of 0.25%. The *y-axis* of Fig. 7(b) shows the overall runtime. This figure demonstrates that the existing algorithm is inefficient when a window size becomes large or number of distinct items in a window is large. It also shows that our algorithm is efficient for handling a large window with a large number of distinct items.

4.4   Memory Efficiency of the HUPMS Algorithm

Prefix-tree-based frequent pattern mining techniques [7], [6], [9], [16], [8] have shown that the memory requirement for the prefix trees is low enough to use the

Table 1 Memory comparison (MB)

Dataset (δ) (window and batch size)	HUPMS	MHUI-TID
T10I4D100K (2%) (W=3B, B=10K)	5.178	38.902
BMS-POS (3%) (W=3B, B=30K)	13.382	97.563
kosarak (5%) (W=4B, B=50K)	42.704	324.435
Chain-store (0.25%) (W=5B, B=100K)	118.56	872.671

gigabyte-range memory now available. We have also handled our tree structure very efficiently and kept it within this memory range. Our prefix-tree structure can represent the useful information in a very compressed form because transactions have many items in common. By utilizing this type of path overlapping (prefix sharing), our tree structure can save memory space. Moreover, our algorithm efficiently discovers HUPs by using a pattern growth approach and therefore generates much less number of candidates compared to the existing algorithms. Table 1 shows that our algorithm significantly outperforms the existing algorithm in memory usage.

5 Conclusions

The main contribution of this paper is to provide a novel algorithm for sliding window-based high utility pattern mining over data streams. Our proposed algorithm, HUPMS, can capture the recent change of knowledge in a data stream adaptively by using a novel tree structure. Our tree structure, HUS-tree, maintains a fixed sort order and batch-by-batch information, therefore, it is easy to construct and maintain during the sliding window-based stream data mining. Moreover, it has the *"build once mine many"* property and efficient for interactive mining. Since our algorithm exploits a pattern growth mining approach, it can easily avoid the problem of the level-wise candidate generation-and-test approach of the existing algorithms. Hence, it significantly reduces the number of candidate patterns as well as the overall runtime. It also saves a huge amount of memory space by keeping the recent information very efficiently in an HUS-tree. Extensive performance analyses show that our algorithm outperforms the existing sliding window-based high utility pattern mining algorithms in both runtime and memory usage, efficient for interactive mining and can handle a large number of distinct items and transactions.

References

1. Agrawal, R., Srikant, R.: Fast Algorithms for Mining Association Rules. In: 20th Int. Conf. on Very Large Data Bases (VLDB), pp. 487–499 (1994)
2. Chang, J.H., Lee, W.S.: estWin: Online data stream mining of recent frequent itemsets by sliding window method. Journal of Information Sciences 31(2), 76–90 (2005)
3. Chu, C.-J., Tseng, V.S., Liang, T.: An Efficient mining for mining temporal high utility itemsets from data streams. Journal of Systems and Software 81, 1105–1117 (2008)
4. Erwin, A., Gopalan, R.P., Achuthan, N.R.: CTU-Mine: An Efficient High Utility Itemset Mining Algorithm Using the Pattern Growth Approach. In: 7th IEEE Int. Conf. on Computer and Information Technology (CIT), pp. 71–76 (2007)
5. Frequent itemset mining dataset repository,
   http://fimi.cs.helsinki.fi/data/
6. Grahne, G., Zhu, J.: Fast Algorithms for frequent itemset mining using FP-Trees. IEEE Transactions on Knowledge and Data Engineering 17(10), 1347–1362 (2005)
7. Han, J., Pei, J., Yin, Y., Mao, R.: Mining frequent patterns without candidate generation: a frequent-pattern tree approach. Data Mining and Knowledge Discovery 8, 53–87 (2004)

8. Leung, C.K.-S., Khan, Q.I.: DSTree: A Tree structure for the mining of frequent Sets from Data Streams. In: 6th IEEE Int. Conf. on Data Mining (ICDM), pp. 928–932 (2006)
9. Leung, C.K.-S., Khan, Q.I., Li, Z., Hoque, T.: CanTree: a canonical-order tree for incremental frequent-pattern mining. Knowledge and Information Systems 11(3), 287–311 (2007)
10. Li, H.-F., Huang, H.-Y., Chen, Y.-C., Liu, Y.-J., Lee, S.-Y.: Fast and Memory Efficient Mining of High Utility Itemsets in Data Streams. In: 8th IEEE Int. Conf. on Data Mining, ICDM, pp. 881–886 (2008)
11. Li, J., Maier, D., Tuftel, K., Papadimos, V., Tucker, P.A.: No pane, no gain: efficient evaluation of sliding-window aggregates over data streams. SIGMOD Record 34(1), 39–44 (2005)
12. Li, Y.-C., Yeh, J.-S., Chang, C.-C.: Isolated items discarding strategy for discovering high utility itemsets. Data & Knowledge Engineering 64, 198–217 (2008)
13. Lin, C.-H., Chiu, D.-Y., Wu, Y.-H., Chen, A.L.P.: Mining frequent itemsets from data streams with a time-sensitive sliding window. In: 5th SIAM Int. Conf. on Data Mining, pp. 68–79 (2005)
14. Liu, Y., Liao, W.-K., Choudhary, A.: A Two Phase algorithm for fast discovery of High Utility of Itemsets. In: Ho, T.-B., Cheung, D., Liu, H. (eds.) PAKDD 2005. LNCS (LNAI), vol. 3518, pp. 689–695. Springer, Heidelberg (2005)
15. Pisharath, J., Liu, Y., Parhi, J., Liao, W.-K., Choudhary, A., Memik, G.: NU-MineBench version 2.0 source code and datasets, http://cucis.ece.northwestern.edu/projects/DMS/MineBench.html
16. Tanbeer, S.K., Ahmed, C.F., Jeong, B.-S., Lee, Y.-K.: Efficient single-pass frequent pattern mining using a prefix-tree. Information Sciences 179(5), 559–583 (2009)
17. Tseng, V.S., Chu, C.-J., Liang., T.: Efficient mining of temporal high utility itemsets from data streams. In: 2nd ACM Int'l. Workshop on Utility-Based Data Mining, UBDM (2006)
18. Yao, H., Hamilton, H.J.: Mining itemset utilities from transaction databases. Data & Knowledge Engineering 59, 603–626 (2006)
19. Yao, H., Hamilton, H.J., Butz, C.J.: A Foundational Approach to Mining Itemset Utilities from Databases. In: 4th SIAM Int. Conf. on Data Mining, pp. 482–486 (2004)
20. Zheng, Z., Kohavi, R., Mason, L.: Real world performance of association rule algorithms. In: 7th ACM SIGKDD, pp. 401–406 (2001)

Resolving Sluices in Urdu

Mohammad Abid Khan, Alamgir Khan, and Mushtaq Ali

Abstract. This research paper presents a rule based algorithm for the resolution of sluices occurring in dialog context. In this algorithm, the syntactic approach is followed for the resolution of sluices. Real text of Urdu language from different sources such as books, novels, magazines and web resources is taken into account for the testing of this algorithm. The algorithm achieved accuracy of 78%.

Keywords: sluicing, Wh-words, Ellipses, Resolution.

1 Introduction

When any Wh-word occurs at the end of an English sentence, usually some information is made elided after the Wh-words, though it is not always essential but is due to the custom of the language and for the purpose of briefness. For human beings (native speakers or those who understand a language), it is easy to make sense of the constituents that are missing from the sluicing construction. But in order for a language to be processed by a machine, there is a need of some explicit intelligence to be brought into play.

1.1 Sluicing

"Sluicing is the Ellipsis Phenomenon in which the sentential portion of a constituent question is elided leaving only a Wh-Phrase remnant" [1]. Consider the following examples of sluicing in English language.

Jack bought something, but I don't know what?
Beth was there, but you'll never guess who else?
A: Someone called.   B: Really? Who?

In all the above examples, the Wh-word at the end of the sentence makes them sluicing. The first two examples represent sluicing in mono-sentential case while the third example is a matrix sluicing [2] i.e. sluicing occurs both in monolog and

Mohammad Abid Khan, Alamgir Khan, and Mushtaq Ali
Department of Computer Science,
University of Peshawar, Pakistan
e-mail: mabid@upesh.edu.pk

Roger Lee (Ed.): SNPD 2010, SCI 295, pp. 115–126, 2010.
springerlink.com   © Springer-Verlag Berlin Heidelberg 2010

dialog. In English, the Wh-words are "When", "What", "Where", "Which", "Who", "How" and "Why". Also other words can be used to inquire about specific information such as "Which one", "Whom", "Whose", "How much", "How Many", "How long" and "What Kind (of)" [3]. Roughly speaking, it is possible to elide everything in an embedded question except the question word and a stranded preposition if there is a sentence earlier in the discourse that duplicates the meaning of the question called matrix sluicing [2]. Sluicing is a widely attested elliptical construction in which the sentential part of a constituent question is elided, leaving only the Wh-phrase. The full ranges of Wh-phrases, found in constituent questions, are found in sluicing [10, 11, 12]. Chung and co-authors further divide sluicing into two types [11]. In the first type, the sluiced Wh-phrase corresponds to some phrase in the antecedent sentence. The second type they name the 'sprouting', in which no overt correlate is found [4]. Ginzburg and Sag distinguish between direct and reprise sluices [13]. For direct sluicing, the question is a polar question, where it is required to be a quantified proposition. The resolution of the direct sluice consists in constructing a Wh-question by a process that replaces the quantification with a simple abstraction.

1.2   Resolving Ellipses in Sluicing

For complete interpretation of matrix sluicing, it is necessary for each of these sentences to have presented all of the required constituents/phrases in them. As in sluicing, the information after the Wh-word is missing in many cases and for a language to be processed by machine/computer, this missing information is needed to be recovered for effective treatment. Therefore, the material missing after the Wh-words in sluicing would be recovered from the text within the context of this sluicing. As here matrix sluicing is considered, therefore the information for resolution of ellipses would be looked in the antecedent clause of the matrix sluicing. The antecedent clause is the one from where information is recovered in the sluiced clause while sluiced clause is the one in which sluicing occurs.

The importance of ellipses resolution in sluicing comes from the fact that, for nearly all of the applications of NLP, sentences need to be in the form of simple text so as to be properly processed.

In this work, the published text of Urdu language from various genres such as books, newspapers, magazines and internet is searched out for the presence of sluicing in it and various theoretical and computational aspects of sluicing are explored within this real text. After this, some steps are taken to devise a procedure for resolving ellipses in sluicing. There has been no computational work on sluicing resolution in Urdu from the syntactic point of view. Here a syntactic approach for the resolution of sluicing is presented. Section-2 describes various factors involved in the resolution of sluicing in Urdu. Section-3 provides an algorithm for the computational resolution of sluicing in Urdu language and in section-4 evaluation and error analysis of the suggested approach is provided.

2 Sluicing in Urdu

Urdu Historically Spelled Ordu is a central Indo-Aryan Language of the Indo-Iranian branch, belonging to the Indo-European family of languages. Standard Urdu has approximately the twentieth largest population of native speakers, among all languages. According to the SIL Ethnologue (1999 data), Urdu اردو/Hindi is the fifth most spoken language in the world. According to George Weber's article "Top Languages: The World's 10 Most Influential Languages in Language Today", Hindi/Urdu is the fourth most spoken language in the world, with 4.7 percent of the world's population, after Mandarin, English, and Spanish [7]. With regards to word order, Hindi-Urdu is an SOV (Subject, Object and Verb) language.

Whether or not you would expect to find sluicing in Hindi-Urdu is depending on your account of sluicing. In fact, sluicing is generally possible in Hindi-Urdu [14, 15]. The Wh-words used in Urdu are "کون" [kʊn] (Who), "کیسے"[kəseI] (How), "کیوں" [kəʊñ] (Why), "کیا" [kəya:] (What),"کب" [kɒb] (When), "کہاں\کدھر" [kədhɒr]/[kɒhɒñ] (Where), "کونسا" [kʊnSɒ] (Which one), "کس کا" [KəsKɒ] "(whose), "کس کو" [KəsKʊ] (Whom) and "کتنا" [kt̪nɒ:] (how much).

The purpose of this paper is to discuss the theoritical aspects of sluicing occuring in Urdu with an attempt to make ground for resolution of ellipses in these sluicing constructions. Sluicing when occur in Urdu has different forms. Most of these are discussed in this paper.

Example 1:   A: وہ تیرنا جانتے ہیں
                    [hæñ] [dʒɒnt^eI] [t^ərnɒ:] [wʊh]
                    [are][know][swimming][They]
                    They know swimming.
                              B: کون؟
                              [kʊn]
                              Who?
                    (نسیم حجازی،"انسان اور دیوتا"،صفحہ نمبر16)

In Example1, in the sluiced clause the Wh-word "کون" [kʊn] (Who) exists and this "کون " [kʊn] (Who) makes the sentence a sluicing. In order to make sense of this sluiced clause, there is a need to recover the missing information. Usually, this missing information which is caused due to sluicing is present there before or after the sluicing. In order to resolve this ellipsis, the already existing information may be used in some mechanism to produce the complete sentence. Here in the sluiced clause, the phrase (usually verb phrase) is missing [8]. This phrase may be recovered from the antecedent clause to resolve the ambiguity and present proper interpretation of the text. So the sluicing sentence after ellipses resolution becomes:

کون تیرنا جانتے ہیں؟

[hæñ] [dʒɒnt^eI] [t^ərnɒ:] [kʊn]
[are] [knows] [Swimming] [Who]
Who knows swimming?

Example 2:   A: ماتا!بارش کہاں سے آتی ہے؟
[heI] [a:t^ə] [seI] [kɒhɒñ] [bɒ(r)əʃ] [mɒt^ɒ]
[is] [comes] [from] [where] [rain] [god]
god! From where rain comes?

B: بادلوں سے اور کہاں سے؟
[seI] [kɒhɒñ] [ɒʊ(r)] [seI] [bɒdlʊñ]
[from] [where] [else] [from] [clouds]
From clouds, from where else?

(نسیم حجازی،"انسان اور دیوتا"،صفحہ نمبر169)

In Example 2, in the sluiced clause there exists some phrases as well before the
Wh-word but still after the Wh-phrase   "کہاں سے" [kɒhɒñ] [seI] (from where)
there is an ellipsis i.e. the verb phrase " آتی ہے " [a:t^ə] [heI] (comes) of the source
sentence is missing in the sluiced clause and it should be recovered from the
antecedent clause which is again a verb phrase. This verb phrase should be
inserted into the sluiced clause after the Wh-phrase to have the complete and
meaningful sense of the sentence. In most of the cases in sluicing, the verb phrase
remains absent after the Wh-phrase in the sluiced clause. In ellipses resolution, in
Wh-construction, the focus comes implicitly to answering the question [9].
However, here in resolving ellipses in sluicing, there is no intention of providing
answer to the Wh-word.

   Whenever in the sluiced clause a preposition comes with the Wh-word forming
a prepostional phrase, the nature of sluicing changes. This phenomenon due to
Wh-prepostional phrase is termed as swipping[1]. Swipping is defined as"
Monomorphemic Wh-words selected by certain (generaly "simplex") prepositions,
can undergo a surprising inversion in sluicing" [1].

Example 3:   A: اب اپنی غلطی کا خمیازہ بھگتو؟
[bʰʊghtʊ] [khɒmyɒzɒ] [kɒ] [gɒlt^i] [ɒpni] [ɒb]
[suffer] [pay] [of] [blunder] [your] [now]
Now pay the penalty of your mistake.

B: کب تک؟
[t^ɒk] [kɒb]
[upto] [when]
Upto what time?

(محمد سعید،" ،"مدینتہ الیہود"جلد اوّل،1960،صفحہ نمبر72)

Here in Example 3, when in the sluiced clause, only the Wh-word   "کب" [kɒb]
(When) exists then the interpretation of this sluicing sentence is completely
different from the case when there is   "تک " [t^ɒk] (upto) the preposition
immediately after this Wh-word.

Example 4:   A: میں کچھ اور سننا چاہتا ہوں ۔
[hʊñ] [ʧa:ht^ɒ] [sʊn:nɒ] [ɒʊ(r)] [kʊ ʧʰ] [mæñ]
[want] [listen] [else] [some] [I]
I want to listen something else.

B: کیا؟
[kəya:]
What?

(محمد سعید،" ،"مدینتہ الیہود"جلد اوّل،1960،صفحہ نمبر 86)

In Example 4, in the sluiced clause, there is a lexical Wh-word "کیا" [kəya:] (what). To recover this ellipsis due to the sluicing when the verb phrase "چاہتا ہوں" [ʧa:ht^ɒ][hʊñ](want, 1st person) is to be inserted after "کیا" [kəya:] (What), there is a need to transform the auxiliary verb ہوں (to be, 1st person singular) into" ہو "(to be, 2nd person singular). Similarly, "چاہتا" will be transformed to "چاہتے ". The verb in this example is in the present tense. Similar is the case with the future tense as is shown in the following example.

Example 5:   ۔میں تمھیں یہ مشورہ ہر گز نہیں دوں گا :A

[ga] [dʊn] [nah^i] [g^iz] [he(r)] [mɒshwɒrɒ] [yɒh] [tumhæñ] [mæñ]
[will] [give] [never] [suggestion] [this] [you] [I]
I will never give you this suggestion.

کیوں؟:B
[kəʊñ]
Why?

(محمد سعید،" ،"مدینتہ الیہود"جلد اوّل،1960،صفحہ نمبر88)

In Example 5, the antecedent clause is in future tense and is first person singular. To resolve this sluicing sentence, there is a need to insert the verb phrase from the antecedent clause after the Wh-word " کیوں" [kəʊñ] (Why) in the sluiced clause. Also, the verb along with the auxilary needs to be converted into a form having the verb ending so as it agrees to the subject of the sentence which should be in 2nd person now.

Example 6:   ۔کل صبح سمندر کے کنارے مقابلہ ہے :A
[heI] [mʊkɒbla:] [kəna:(r)eI] [keI] [sɒmɒndɒ(r)] [sʊbhɒ] [kɒl]
 [is] [competition] [side] [on] [sea] [morning] [tommorrow]
Tommorrow there is a competition on sea side.

کس چیز کا؟:B
[kɒ] [ʧəz] [Kəs]
[of][thing] [what]
Of what?

(محمد سعید،" مدینتہ الیہود"جلد اوّل،1960، صفحہ نمبر 14)

In Example 6, the nominal Wh-phrase"کس چیز کا " [Kəs][ʧəz][kɒ] (of what) exists in the sluiced clause. It inquires about something and needs the sluicing part to be recovered from ellipsis that it has. When there is no proper verb existing in the antecedent clause, the noun "مقابلہ " [mʊkɒbla:](competition) along with the copula verb"ہے"[heI] (is) is to be inserted after the Wh-phrase "کس چیز کا " [Kəs][ʧəz][kɒ] (of what) in the sluiced clause so that the ellipsis-resolved sentence B becomes "کس چیز کا مقابلہ ہے؟"[Kəs] [ʧəz] [kɒ] [mʊkɒbla:] [heI] (competition of what thing).

Example 7:   اسے صبح سویرےروانہ ہونا تھا :A
[t^ha:] [hʊna:] [(r)ɒwɒna:] [sɒwə(r)eI] [sʊbhɒ] [əseI]
             [was] [by] [leave] [early] [morning] [he]
He was supposed to leave early in the morning.

کہاں؟:B
[kɒhɒñ]
Where?

(محمد سعید،" مدینتہ الیہود"جلد اوّل، 1960، صفحہ نمبر 35)

Example 8:   A: یہ اب باہر جا رہی ہیں

[hæñ] [rɒhi] [ʤɒ] [bɒhə(r)] [ɒb] [yɒh]
[are] [going] [outside] [now] [they]
These (female) are now going outside.

B: کہاں؟

[kɒhɒñ]

Where?

(محمد سعید،" مدینتہ الیہود"جلد اوّل،1960، صفحہ نمبر 39)

Example 9:   A: لیکن بلی کو ڈھونڈھنے کو جی نہں چاہتا۔

[ʧa:ht^ɒ] [nɒhiñ] [ʤi] [kʊ] [dʰʊñdneI] [bəl:li] [ləkən]

[do not] [want] [to] [search] [cat] [but]

But I do not want to search the cat.

B: کیوں؟

[kəʊñ]

Why?

(محمد سعید،" مدینتہ الیہود"جلد اوّل،1960، صفحہ نمبر 38)

In Examples 7, 8 and 9, the verb phrases " روانہ ہونا تھا"[(r)ɒwɒna:][hʊna:][t^ha:]
(supposed to leave) , " جا رہی ہیں" [ʤɒ] [rɒhi] [hæñ] (going) and "جی نہں چاہتا "
[ʤi] [nɒhiñ][ʧa:ht^ɒ] (do not want) are to be inserted respectively after the Wh-
words in the sluiced clauses, to resolve sluicing. In most of the examples, the
analysis is such that to resolve the sluicing and disambiguate the sentence, the
verb phrase in the antecedent clause should be inserted into the sluiced clause after
the Wh-word/Wh-phrase.

After analysing the examples, it is concluded that the verb agrees with the
subject. In the resolution of sluicing, appropriate verb ending is placed with each
verb based on its agreement with the subject. These appropriate verb endings
should be used in the sluiced clause after the Wh-word/Wh-phrase according to
Table-I, II and III given in appendix. The abbreviations used are given in Table-
IV. The verb will be transformed according to the subject from the antecedent
clause and will be placed after the Wh-word in the sluiced clause. Similarly, the
auxiliaries also change accordingly with the subject in the resolution of sluicing.
The transformation rules for present, past and future tenses are also given in
Table-I, II and III respectively. Similarly, the tense after resolution of sluicing
will remain the same as in the antecedent clause as has been noticed in all the
examples that the tense remains the same before and after the resolution of
sluicing. Also the rules are applied to both the voices i.e. active or passive voice.
In both the cases, the same rules are applied. The voice does not matter any
change in the rules. Also, in the feminine as observed in Examples 2 and 8 where
after resolution the verb-phrases" آتی ہے " [a:t^ə] [heI] (comes) and "جا رہی ہیں"
[ʤɒ] [rɒhi] [hæñ] (going) are elided after the Wh-word in the sluiced clause
hence no change has been observed in the transformation of verbs. So we ignore
the case of feminine in the tranformation of verb endings while in case of
auxiliaries we have specified the rules where necessary.

Sluicing occurs in almost each and every language [1]. This research work is
based on the literature survey of sluicing occuring in English, Urdu and Pushto
languages. It has been observed that sluicing occuring in Urdu language is almost

identical to that occuring in English and Pushto. There has been no computational work on sluicing resolution in Pushto so far, although very little work has been done on Ellipses resolution [9]. We have worked out examples from real text including books, novels and online sources to find out the fact that sluicing occuring in Urdu is identical to that occuring in Pashto language. Similarly, the work of Jason Merchant shows that sluicing occuring in dialogues in English language is the same in nature to that of sluiicng in Urdu [1].

3 Algorithm

This algorithm resolves sluicing occurring in Urdu language where the antecedent and the sluiced clauses are at a distance of one sentence from each other. This algorithm uses the existing syntactic information for the resolution of sluicing in Urdu language. The antecedent and the sluiced clauses are first tagged and the focus Wh-word in the sluiced clause is identified. Then the corresponding antecedent clause is checked and the verb phrase from the antecedent clause is placed accordingly at the position after the Wh-word/Wh-phrase. The tense and the pronouns are changed according to the rules mentioned in Tables I, II and III.

This algorithm shows accuracy of 78%. It is the first ever rule based algorithm that uses only the syntactic information for the resolution of sluicing in Urdu language. Due to this novelty, it shows better results than those algorithms where both syntactic and semantic information are taken into consideration. Although at present there are algorithms which show accuracy greater than this algorithm but they are specific to that task or language. When we tested our examples on already existing algorithms the results were unsatisfactory and we had to go for this solution. During literature survey, the algorithm presented by Sobha and Patnaik was also tried for good results but their approach to the resolution is totally different than our solution [16]. Their rules are not applicable on our data and for this very reason we decided to go for new solution. In future, the capability of this algorithm will further increase as it is just a prototype at present. The algorithm is:

1. Tag both the antecedent and sluiced clauses.
2. Identify noun and verb phrases in both the antecedent and sluiced clauses.
3. Identify the Wh-word/ phrase in the sluiced clause.
4. Check the tense, person and number of the verb and auxilary (if any) of the antecedent sentence:

    If 1st person

    > Convert the verb (or auxilary if any) of the antecedent clause by changing the verb ending (and/or auxilary) for 2nd person according to its agreement with the subject.

    If 2nd person

    > Convert the verb (or auxilary if any) of the antecedent clause by changing the verb ending (and/or auxilary) for 1st person according to its agreement with the subject.

    There is no need to transform the verb (or auxilary). The same verb endings (or auxilary if any) will be used in the sluiced clause.

5. Remove the sign of interogation from the sluiced clause.
6. Now place the verb phrase from the antecedent clause in the sluiced clause after the Wh-word/Wh-phrase to resolve the sluicing.
7. Put the sign of interogation at the end of this sentence which is removed in step 6.

First of all, both the clauses i.e antecedent and sluiced clauses were manually tagged as no tagger for Urdu language is available for automatic tagging. In the second step, noun and verb phrases were identified in both the sentences. In the 3rd step the Wh-word/phrase is identified in the sluiced sentence. Step 4 checks the tense, person and number of the verb and auxilary in the antecedent sentence: if the verb ending or auxilary shows that it is for first person, transform the verb or auxilary of the antecedent clause according to Table-I, II and III into second person depending on the tense, person and number of the verb and its agreement with the subject. Nevertheless, if the verb ending or auxilary is for the second person, convert it into first person as per Table-I, II and III depending on the tense, person and number and its agreement with the subject. Similarly in step 5, the sign of interrogation is removed from the sluiced clause and in step 6 the verb phrase from the antecedent clause is placed in the sluiced clause after the Wh-word/Wh-phrase to resolve ellipsis in the sluicing. Finally in step 7, the sign of interrogation is added at the end of the resolved sentence.

4   Evaluation and Error Analysis

The Algorithm was tested on Urdu text examples which were manually tagged. Hundreds of examples from published text of Urdu language from various genres such as books, newspapers, magazines and internet were taken for testing. In these examples, pairs of sentences containing different Wh-words are taken. Some have multiple Wh-words while in some cases the Wh-words fall several sentences back somewhere in the context. It was analyzed that when the antecedent and sluiced clauses occur in pairs the algorithm works efficiently but when multiple Wh-words occur somewhere back in several previous sentences in the context then it becomes difficult to resolve the sluices which may be recovered by defining more rules and enhancing the algorithm and is left for future work. However, beside these discrepancies, a success rate of 78% is achieved. The errors, which are 22% of the tested text, are due to multiple Wh-words in the sluiced clause and referring of Wh-word to the antecedent clause several sentences back. For this purpose, analysis of Urdu text has been carried out and the distances from the sluiced clause and antecedent clause are calculated on the basis of number of sentences between them which is shown percentage wise in Table-V. For this purpose different novels and books are consulted and the distances are recorded accordingly. Here the fact can be noticed that mostly the Wh-clause occurred alternately with the antecedent clause as the most frequently occuring distance is 1, followed by 2, 3 and so on. A total of 45.5% errors occur due referring of Wh-word to the antecedent clause several sentences back while the remaining 54.5% occur due to confrontation of multiple Wh-words in the sluiced clause in which it is difficult to find out which Wh-word is a focus word. At present, it is not possible to incorporate these in the algorithm as it needs a lot of work to be done and need more time for successful solution, which is left for future work. A simple mechanism for ellipses resolution in ellipses occuring in the sluicing part of these sentences was presented. Here our focus was mainly on the action i.e recovery of verb-phrase.

References

1. Merchant, J.: Sluicing. In: Syntax Companion Case 98, University of Chicago (2003)
2. Kim, J.: Sluicing. Submitted to Department of Linguistics, University of Connecticut (1997) (in English)
3. Lappin, S., Ginzburg, J., Fernandez, R.: Classifying Non-Sentential Utterances in Dialogue: A Machine Learning Approach. Computational Linguistics 33(3), 397–427 (2007)
4. Merchant, J.: Islands and LF-movement in Greek sluicing. Journal of Greek Linguistics 1(1), 41–46 (2001)
5. Rahi, A.: Qaitba Bin Muslim, 1st edn., Lahore (1997)
6. Saeed, M.: Madina-tul-aihood aik tarih aik naval, 1st edn., Lahore (1960)
7. Weber's, G.: Top Languages. Language Monthly 3, 12–18 (1997)
8. Sajjad, H.: Urdu Part of Speech Tagset (2007), http://www.crulp.org/software/langproc/POStagset.htm (accessed March 2, 2009)
9. Ali, et al.: Ellipses Resolution in wh-constructions in Pashto. In: Proceedings of IEEE International Multitopic Conference (2008)
10. Ross, J.R.: Guess who? Papers from the 5th Regional Meeting of Chicago Linguistic Society, pp. 252–286 (1969)
11. Chung, S., Ladusaw, W., McCloskey, J.: Sluicing and Logical Form. Natural Language Semantics (1995)
12. Levin, L.: Sluicing: a lexical interpretation procedure. In: Bresnan, J. (ed.) The mental representation of grammatical relations. MIT Press, Cambridge (1982)
13. Ginzburg, J., Sag, I.: Interrogative Investigations. CSLI Publications, Stanford (2001)
14. Mahajan, Anoop.: Gapping and Sluicing in Hindi. GLOW in Asia, New Delhi (2005)
15. Hijazi, N.: Insan aur dewta, 1st edn., Qomi Kutab Khana Feroz Pur Lahore
16. Sobha, L., Patnaik, B.N.: VASISTH-An Ellipses Resolution Algorithm for Indian Languages. In: An international conference MT 2000: machine translation and multilingual applications in the new millenium. University of Exeter, British computer society, London (2000)

Appendix

Table 1 Transformation of Verb-Endings and Auxiliaries (Present Tense)

	Tense	Person	Number	Masc.		Auxiliaries	
				Verb Ending	Transformation	Auxiliary Ending	Transformation
Present Tense	Indefinite	P-1	sing	ڈ/وں	ڈ / ے	وں	و
			Plu	ں/ے	ں/ے	ں	ں
		p-2	sing	ڈ / ے	ڈ/وں	و	وں
			plu	ں/ے	ں/ے	ں	ں
		p-3	sing	ا	ا	ے	ے
			plu	ے	ے	ں	ں
	Continuous	p-1	sing	ا	ا	رہا ہوں	رہے ہو
			Plu	ا	ا	رہے ہیں	رہے ہیں
		p-2	sing	ا	ا	رہے ہو	رہا ہوں
			Plu	ا	ا	رہے ہیں	رہے ہیں
		p-3	Sing	ا	ا	رہا/رہی ہے	رہا/رہی ہے
			Plu	ا	ا	رہے ہیں	رہے ہیں
	Perfect	p-1	sing	ا	ا	چکا ہوں	چکے ہو
			Plu	ا	ا	چکے ہیں	چکے ہیں
		p-2	sing	ا	ا	چکے ہو	چکا ہوں
			Plu	ا	ا	چکے ہیں	چکے ہیں
		p-3	Sing	ا	ا	چکا/ چکی ہے	چکا/ چکی ہے
			Plu	ا	ا	چکے ہیں	چکے ہیں
	Perfect Continuous	p-1	sing	ا	ے	رہا ہوں	رہے ہو
			Plu	ے	ے	رہے ہیں	رہے ہیں
		p-2	sing	ے	ا	رہا ہو	رہا ہوں
			Plu	ے	ے	رہے ہیں	رہے ہیں
		p-3	Sing	ا	ا	رہا/رہی ہے	رہا/رہی ہے
			Plu	ے	ے	رہے ہیں	رہے ہیں

Table 2 Transformation of Verb-Endings and Auxiliaries (Past Tense)

	Tense	Person	Number	Masc.		Auxiliaries	
				Verb Ending	Transformation	Auxiliary Ending	Transformation
Past Tense	Indefinite	P-1	sing	ا	ے	تھا	تھے
			Plu	ے	ے	تھے	تھے
		p-2	sing	ے	ا	تھے	تھا
			plu	ے	ے	تھے	تھے
		p-3	sing	ا	ا	تھا	تھا
			plu	ے	ے	تھے	تھے
	Continuous	p-1	sing	ا	ا	رہا تھا	رہے تھے
			Plu	ا	ا	رہے تھے	رہے تھے
		p-2	sing	ا	ا	رہے تھے	رہا تھا
			Plu	ا	ا	رہے تھے	رہے تھے
		p-3	sing	ا	ا	رہا/ رہی تھی	رہے تھے
			Plu	ا	ا	رہے تھے	رہے تھے
	Perfect	p-1	sing	ا	ا	چکا تھا	چکے تھے
			Plu	ا	ا	چکے تھے	چکے تھے
		p-2	sing	ا	ا	چکے تھے	چکا تھا
			plu	ا	ا	چکے تھے	چکے تھے
		p-3	sing	ا	ا	چکا/ چکی تھی	چکے تھے
			plu	ا	ا	چکے تھے	چکے تھے
	Perfect Continuous	p-1	sing	ا	ے	رہا تھا	رہے تھے
			Plu	ے	ے	رہے تھے	رہے تھے
		p-2	sing	ے	ا	رہے تھے	رہا تھا
			Plu	ے	ے	رہے تھے	رہے تھے
		p-3	sing	ا	ا	رہا/ رہی تھی	رہا/ رہی تھی
			Plu	ے	ے	رہے تھے	رہے تھے

Table 3 Transformation of Verb-Endings and Auxiliaries (Future Tense)

	Tense	Person	Number	Masc.		Auxiliaries	
				Verb Ending	Transformation	Aux	Transformation
Future Tense	Indefinite	P-1	sing	وں	و	گا	گے
			Plu	و	و	گے	گے
		p-2	sing	و	وں	گے	گا
			plu	و	و	گے	گے
		p-3	sing	ے	ے	گا/گی	گا/گی
			plu	و	و	گے / گی	گے / گی
	Continuous	p-1	sing	ا	ا	رہے ہوں گا	رہے ہوں گے
			Plu	ا	ا	رہے ہوں گے	رہے ہوں گے
		p-2	sing	ا	ا	رہا ہو گا	رہا ہوں گا
			Plu	ا	ا	رہے ہوں گے	رہے ہوں گے
		p-3	sing	ا	ا	رہا/رہی ہو گی	رہا/رہی ہو گی
			Plu	ا	ا	رہے ہوں گے	رہے ہوں گے
	Perfect	p-1	sing	ا	ا	چکا ہوں گا	چکے ہو گے
			Plu	ا	ا	چکے ہوں گے	چکے ہوں گے
		p-2	sing	ا	ا	چکے ہو گے	چکا ہوں گا
			plu	ا	ا	چکے ہوں گے	چکے ہوں گے
		p-3	sing	ا	ا	چکا/ چکی ہو گی	چکا/ چکی ہو گی
			plu	ا	ا	چکے ہوں گے	چکے ہوں گے
	Perfect Continuous	p-1	sing	ا	ے	رہا ہوں گا	رہے ہو گے
			Plu	ے	ے	رہے ہوں گے	رہے ہوں گے
		p-2	sing	ے	ا	رہے ہو گے	رہا ہوں گا
			Plu	ے	ے	رہے ہوں گے	رہے ہوں گے
		p-3	sing	ا	ا	رہا/ رہی ہو گی	رہا/ رہی ہی گی
			Plu	ے	ے	رہے ہوں گے	رہے ہوں گے

Table 4 Table of abbreviations:

Abbreviation	Actual Word	Abbreviation	Actual Word
p-1	1st Person	Masc.	Masculine
p-2	2nd person	Sing	Singular
p-3	3rd person	Plu	Plural

Table 5 Table calculating distances

Distance (sentences)	Total	Percentage
1	210	70.50%
2	39	13.11%
3	12	5.03%
0	9	4.57%
4	7	3.59%
5	1	0.30%
6	1	0.30%
10	1	0.30%

Context Sensitive Gestures

Daniel Demski and Roger Lee

Abstract. Touchscreens and stylus input are becoming increasingly popular in modern computers. However, many programs still require keyboard input in order to take full advantage of their functionality. In order to increase the functionality of screen-based input, gestures are being adopted to input commands. However, current gesture interfaces are not sufficiently context-based, which limits their usefulness to a few applications, and they are not dynamic or interactive, which wastes time and keeps the user from feeling comfortable using them. In order to improve gesture functionality, we used off-the-shelf hardware and implemented a gesture-based system which changed gesture function based on currently running application, and based on recent gesture activity. The result of application-sensitivity was a system which is useable on many modern computers and provides useful functionality in a variety of applications. The results of sensitivity to previous gestures, however, was only minimal interactivity.

Keywords: gesture, interface, context-sensitive.

1 Introduction

The common hardware elements of computer user interfaces are the screen, keyboard and mouse. The mouse is capable of pointing, clicking, dragging,

Daniel Demski
Software Engineering and Information Technology Institute,
Computer Science Department, Central Michigan University
e-mail: demsk1da@cmich.edu

Roger Lee
Software Engineering and Information Technology Institute,
Computer Science Department, Central Michigan University
e-mail: lee1ry@cmich.edu

Roger Lee (Ed.): SNPD 2010, SCI 295, pp. 127–137, 2010.

and perhaps scrolling, capabilities with which it is able to perform complex interactions with software through use of graphical user interface elements such as menus which allow the user to specify commands to run. Keyboards are mainly used for text entry, but most programs also employ keyboard shortcuts to bypass time-consuming menu interactions and allow the user to quickly specify commands. However, many modern devices do not come equipped with a keyboard, robbing the interface of this avenue for receiving commands. Therefore it is desirable to increase the amount of information which can be expressed by using the mouse. One method for accomplishing this is using gestures.

Gestures are mouse movements which the computer recognizes and acts upon, such as a swipe or circle. They are analogous to keyboard shortcuts in that they provide quick access to functionality for a user who knows that the gesture is there, rather than functioning like a menu which users can explore.

Gestures can be useful in conventional user interfaces with a screen, keyboard and mouse, and for this reason they are implemented in some desktop applications such as the Opera web browser [11]. However, the recent increase in use of touch screens, tablet devices, and mobile computers has created the first real need for gesture interaction. Gestures are being deployed in touch devices as an important aspect of the user interface, but these interfaces too often over-generalize the correspondence between gestures and keyboard shortcuts by mapping a single gesture to a single key combination. The present study seeks to improve upon the gestural interface in commercially available tablet devices by demonstrating the usefulness of creating more complex relationships in which one gesture can map to one or more keystrokes depending on the situation.

2 Background and Review of Literature

There are many different aspects which make up gestural interfaces, and correspondingly many different subtopics in the research on the subject. Because of the tight coupling of all these components in an implemented system, however, solutions are rarely applicable independent of the other decisions made within a system.

One of the subtopics in which a fair amount of work has been done is the recognition of the shapes which constitute gestures in a given system. What basic movements should be distinguished between is the topic of papers such as Roudaut et al. 2009 [12]. Most systems simply consider the lines drawn by pointer input (in addition to state information such as modifier keys and mouse buttons), but Roudaut et al. suggest considering how the input was made; more specifically, whether the thumb was rolled or slid across the screen. Another way of considering the manner in which input is produced is to extrapolate and track the full position of the hand, in systems such as that of Westerman et al. 2007 [15]. Such strategies can be accomplished with or without special hardware; Westerman's system, for example, uses proximity sensors to help track the whole hand, but hand tracking can be accomplished using contact information alone. Roudaut et al. determine whether a finger was slid or rolled without using special equipment, but suggest that it could be used.

A more common consideration is the time taken to produce a movement, for example in pen systems which replace a traditional mouse's right-button click with a prolonged click. Hinckley et al. 2005 [?] uses pauses as a component in more complex gestures.

It should be noted that Hinckley et al. distinguish between delimiters, which are gesture components used to separate two parts of a complex gesture, and the the components or building-blocks which make up the gestures.

These building-blocks are the next level up from the movement or input itself. Many systems, of course, need not distinguish between these and the gestures, since gestures are most frequently just composed of a single building-block; but systems with a large set of gestures, such as Unistrokes and other systems for gestural text entry, must either implicitly or explicitly decide which building-blocks to use within single gestures. The question of which gestural building-blocks are possible, which ones are most dependably detectable, and which ones users learn most easily is an interesting one, especially as it relates to the linguistic question of what constitutes a grapheme in written language, and the problem of handwriting or text recognition. Xiang Cao et al., 2007 [2], makes an analysis of how these building-blocks relate to speed of gesture production.

Study of gestures themselves is exemplified by papers like Moyle et al. 2002 [10], which analyzes how best to define and recognize flick gestures, and Roudaut et al. 2009 [12], which (in their experiment 1) considers recognition of 16 different gestures. We were unable to find such analysis for some simple gestures such as circles and Hinckley's pigtail gestures.

The last aspect of the gestures themselves in a gestural interface is the choice of which gestures the system supports. This choice naturally depends greatly on the previous steps, but those steps also depend on this one since gestures must necessarily be recognized differently depending upon which other gestures they contrast with. A rounded stroke is not different from a corner in all systems, nor a straight line from a curve, and this affects both recognition accuracy and the performance accuracy of the user. In other words, the choice of 'building blocks' interacts with the number of gestures needed in the system.

Castellucci et al. 2008 [3] examines the user aspect of this question by comparing different symbol sets for text entry. Wobbrock et al. 2005 [14] looks at methods for improving the guessability of a system's gesture inventory.

Going beyond gestures themselves, there is a question of how gestures relate to the normal system input. For example, in the system of Liao et al. 2005 [7], a button is held to switch to gesture mode, and any input made with the button held is interpreted as a command rather than ordinary pointer input. Typically, in this type of system, mouse movement which constitutes a gesture is not sent to the application, so as to avoid unintentional actions such as drawing, selecting text, or moving text. This of course works differently when the individual application is doing the gesture interpretation, but the principal is the same any input has either a non-gesture effect or a gesture effect, and interpreting an action as both is to be avoided.

This is called a modal or explicit gesture system. The alternative, amodal or implicit gesture recognition, involves detecting gesture shapes within ordinary input. (The terms modal and amodal can also be used to refer to different types of menu systems, as in Grossman et al. 2006 [4].) The appeal of an implicit system is that it is faster and possibly more intuitive for the user (see Li 2005 [8] for an evaluation). We use this approach for our own system.

An implicit approach requires gestures to be markedly different from the normal motions the user makes, and requires gesture recognition not to make the assumption that there is a gesture to be found in input data. Additionally, it can be more difficult to keep gesture input from having unintended effects, because gestures are detected only once they have been completed, or at least started; before that they are treated as non-gesture input. To salvage the situation, a delay can be used like that used to differentiate between single clicks and double clicks; if the start of a gesture is caught within the delay period it is then not sent as ordinary input unless the gesture does not get finished.

Various hybrid techniques exist between fully explicit and implicit systems. Zeleznik et al. 2006 [16] present a system with 'punctuation' to make gestures fairly unambiguously distinguished from normal input. While this is technically an implicit gesture system, and still has ambiguities, it is an interesting and effective approach. Another hybrid approach is to presented in Saund et al. 2003 [13].

The final properties of a gesture system are its scope, and the commands it issues. Many gesture systems, such Scriboli of Hinckley et al. and PapierCraft of Liao et al., are embedded in individual applications and serve only to facilitate interaction with that application. Other gesture systems attempt to be useful system-wide. It is our view that system-wide gesture interfaces are needed, in order to create a consistent overall interface, in order to create more incentive for the user to learn the gesture system, and in order to allow devices without keyboards or mice the same interaction experience as a desktop computer without altering third-party software the user may want.

The commands executed by many gesture system are designed to offer such a replacement. Gesture systems generally replace typing, as in alphabets like Unistrokes, or mouse controls such as the scroll wheel, or keyboard shortcuts. Sometimes, as in the case of the Opera web browser, this is because gestures are perceived as a superior way of accomplishing the input, or a good alternative to use.

3 Objectives

The present study aims to improve upon the gestural input capabilities which come with many currently available tablet device by breaking the one-to-one correspondence between gestures and keystrokes in two different ways. The first way is allowing a gesture to send a different keystroke depending upon the context in which the gesture is made. Thus a gesture might represent Ctrl-Q in one application and Ctrl-X in another, both of which close the current application. This simple change allows gestures to have a consistent behavior across applications. The second improvement is associating multiple key presses to one gesture. For example the user can make a circle gesture to scroll down, and then continue the circle gesture to continue scrolling down rather than repeating a gesture over and over in order to send individual events. This allows a more immediate, dynamic scrolling experience.

The improvements are demonstrated on currently available hardware, implemented in a way which could be repeated by any application wishing to improve its gesture support.

4 Methods

Implementation of context-sensitive and interactive gestures is performed on an HP Touchsmart TX2, a device which us to test the gestures with capacitive touchpad mouse, capacitive touchscreen, and pressure-sensitive stylus input methods. The software used was the Windows Vista operating system with which the device was commercially available, and the Python programming language, which enabled us to take an approach which could be easily implemented in many other operating systems.

The TX2 comes with a customizable gesture interface system which allows the user to associate key combinations with eight 'flick' gestures (in eight directions). For example, a downwards flick could be associated with the page down key, an upwards one with the page up key; leftwards with Ctrl+W, which closes a page in the Firefox Web browser, and rightwards with Ctrl+T, which opens a blank tab. We compare the functionality of this system of fixed gesture-to-keystroke mappings with our own.

The gesture chosen for this implementation was the circle gesture. Concentrating on a single gesture was done only for the sake of simplification. The gesture does not compete with the built-in flick-based system. The implemented system checks for counterclockwise and clockwise circles as different gestures.

The system is implemented as a Python script which records the last second of mouse movement watching for circles. Besides differentiating between clockwise and counterclockwise circles, it checks whether the mouse button was held down during the circle movement. If not, the gesture represents 'open' (clockwise) and 'close' (counterclockwise), and the operations are implemented as context-sensitive key combinations. If a mouse button is down, the gestures are used for scrolling downwards (counterclockwise) and upwards (clockwise). This behavior demonstrates interactive gestures by scrolling incrementally as the user continues the gesture.

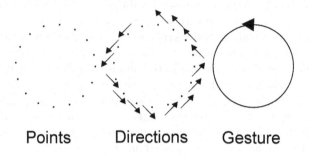

Points   Directions   Gesture

Recorded mouse movements were classified by the direction of movement as compared with the previous mouse position. The gesture is recognized as a series of directional mouse movements. A circle is a gesture which includes all the possible mouse movement directions arranged either in clockwise order or counterclockwise order. Circles were recognized only when they began being drawn at the top, since people would likely feel comfortable beginning the circle the same way every time.

The context-sensitive gestures are implemented by checking the title of the currently selected window, a method which can determine the currently running application, or, if necessary, other information made available through the title bar. Internally, the program has a list of recognition patterns which it steps through one at a time until it recognizes a context with which it has an associated key sequence. If the open window does not match any of these profiles, a default, 'best guess' key sequence can be sent.

The dynamic scrolling begins after one circle gesture with mouse down has been detected. After this, the program switches modes, sending a scroll event each time the gesture is continued further. Our Python implementation uses an arrow keypress event as the scroll event, but a scroll wheel event or a context-sensitive choice between the two would work as well. As the user continues the circle gesture, the program captures all mouse events, with the effect that the mouse pointer on the screen stops moving. This provides the user with extra feedback as to what is going on, and keeps the mouse from performing other, unintended actions as the user uses it to gesture. The program then sends the scroll event periodically while watching for the circle gesture to end. Once the gesture ends mouse control is returned to the user. If instead the user begins to circle in the opposite direction, the scrolling will reverse direction. This dynamic element allows the user faster control over scrolling than static scroll gestures.

The comparison between the context-sensitive, interactive system and the built-in system was performed on several tasks, involving scrolling and opening or closing, in several different applications, and using the mouse, capacitive touchscreen input, and stylus screen input. Comparison is made based on how easily each task can be accomplished.

5 Algorithm

The algorithm described above can be summarized as follows.

Algorithm
1.   **if** mouse moved
2.   Forget events older than max gesture length
3.   Note direction of mouse movement
4.   Add directional movement event to recent events
5.   **if** interactive gesture is most recent gesture event
6.   Check whether mouse movement continues interactive gesture
7.   **if** it does
8.   **if** mouse movement changes direction of interactive gesture
9.   Change direction of interaction
10.   Note direction change as new gesture event
11.   Look up context
12.   Perform keystroke for current gesture and context
13.   **if** maximum number of switches met within recent events
14.   End gesture
15.   Clear recent events
16.   **if** mouse movement does not continue interactive gesture
17.   End gesture
18.   Clear recent events
19.   **if** interactive gesture is not in recent events
20.   Search for gestural pattern in recent movement events
21.   **if** gesture occurred
22.   Look up context
23.   Perform keystroke for current gesture and context
24.   Clear recent events
25.   Note gesture type as recent gesture event

Note, maximum gesture length is a constant; our code used one second. Maximum switches is also a constant; we used five. The gesture search algorithm in our case checked for circles, as follows:

Algorithm *Seek Circles*
1.   **if** all four directions of movement occur in recent movement events
2.   Note first occurrence of each
3.   Note last occurrence of each
4.   **if** first and last occurrences are in clockwise order
5.   Clockwise circle gesture has occurred
6.   **else** **if** first and last occurrences are in counterclockwise order
7.   Counterclockwise circle gesture has occurred

At present, the context is evaluated based on the name of the application which has focus. This allows keyboard shortcuts to be customized to individual applications. Further context detail could potentially be useful, for example if different operations were desirable during text entry; but individual applications form the chief user-interface divide we wish to bridge.

If the application is not recognized, best-guess key combinations are hit, which is similar to what a user might do in an unfamiliar application.

6   Results

The Python script implementation under Windows Vista accomplished the goals of providing contextual, interactive gestures for several common actions.

Scrolling, as compared with the built-in system, was favorable. In the built-in system, a flick gesture could performed the equivalent of a page down or page up. This was efficient for moving one or two pages up or down in the text, but more finely-tuned scrolling could only be performed using the traditional scroll bar interface, which was difficult when using finger input, which is less exact and has trouble with small buttons. Also problematic was scrolling around large documents, in which case using the stylus on a scroll bar could scroll too fast. Flick gestures did not work on with the mouse, only with the screen; mouse-based scrolling was accomplished with a circular motion, making the interface inconsistent.

Scrolling with the interactive gesture system was easier than making a page-down gesture over and over when paging long distances, and scrolling could be sped up by gesturing faster, and slowed down by gesturing slower. It was easy to reverse scrolling direction, and there was no need to resort to the scroll bar. The gesture also worked regardless of application.

Problems included that the scrolling was, at times, slow, which is only a matter of calibration. However, a deeper problem was that the finger input always depresses the mouse, so it could not be used to make mouse-up circles.

The opening and closing gesture, demonstrating context-sensitivity, was also successful. The built-in system could perform the operation in individual applications so long as it was programmed with the correct key sequence, but in applications, such as Emacs, which have longer key sequences for opening a new window (Ctrl-x+5+2) or closing one (Ctrl-x+5+0), it could not do anything. If the system was calibrated for the wrong program and the user made the gesture, the wrong key sequence could be sent, resulting in potentially random results. For example, when the key combinations are calibrated for a Web browser, the 'open' gesture performed in Emacs will transpose two adjacent characters around the cursor, obviously an undesirable and probably a confusing effect. Moreover the user is not likely to change system settings every time they switch applications, so gestures are only likely to work with one application at once.

The dynamic gesture system was able to work with each application included in its lookup table. The 'open' gesture, a clockwise circle without mouse button depressed, was successfully mapped to Ctrl-T in Firefox, Ctrl-N in several other applications, and the sequence Ctrl-X+5+2 in Emacs. A 'guess' for unfamiliar applications was set as Alt-F+Alt-X+Alt-C, which in many Windows applications opens the File menu, then tries to Close or eXit.

There were applications, however, which could not be included in this scheme. Chrome, the Web browser created by Google, does not have window title bar text available to the Windows API, so it could not be identified, and the default key sequence did not work with it. However, the system still was more useful and consistent than the built-in gestures, and perhaps it could be extended to understand more applications.

7 Analysis of Results

The system achieved a consistent interface across more applications than the system to which it was compared, and a more dynamic, interactive sort of gesture. A consistent interface allows the user to act instinctively, and with support for many applications, this instinctive act becomes much more useful. Interactive gestures, too, encourage the user with immediate feedback. The overall effect is increased user confidence and more efficient use of the gesture system.

However, the approach of using window title text to differentiate context, though often adequate, has limits. Applications which do not provide title bar text exist, as well as applications which do not state their identity within the title bar.

Using mouse button state to control the gesture behavior was limiting as well, though this is only a small concern since in a more complete system there would be more gestures.

8 Further Study

In general, gesture recognition cannot be left up to the operating system or device drivers, because different devices have different needs. The approach in this paper has been to use the same gestures for similar behavior across different applications by specifying how that behavior is implemented on a per-application basis. However, for more advanced behaviors, the application itself must process the input. This can be seen in multi-touch platforms such as UITouch [1] and Lux [9], where all points of contact with the screen are made available to an application. This stands in contrast to the approach of recognizing a few multi-touch gestures such as zooming at the operating system or driver level, which loses much of the detail which gives this sort of interface so much potential.

Thus, further research in the direction of making gestures more useful should include a concentration on specific applications which the present study is missing. For example, an application for manipulating 3-D models might require gestures which specify which axis a mouse movement affects. There are several aspects of interactivity and context-sensitivity relevant to such situations. The relevance of context sensitivity is that it is best if the user can guess what gestures to make, so familiar gestures such as those for horizontal scrolling, vertical scrolling, and zooming might be appropriate if they are available to be borrowed. Interactivity is especially important for manipulation of 3-D objects, and a method which allows the user to quickly and intuitively manipulate the object should be used.

Another important avenue of further study is the circle gesture which was implemented in the present work. Improving recognition of circle gestures was not a goal for our system, but were relevant to its usability. Empirical studies such as that in [10] investigate how people tend to make 'flick' gestures and the implications for recognition of this gesture, but we found no such analysis of the circle gesture, outside of Internet discussions and blogs [6]. Important questions of usability for this gesture include whether users are more comfortable in a system sensitive to the starting point of the circle (top or bottom) and the radius, which can decrease false positives in gesture recognition but make the gesture harder to produce.

References

1. Apple   Developer   Network,   http://developer.apple.com/iPhone/
   library/documentation/UIKit/Reference/UITouch_Class/
   Reference/Reference.html (Cited March 27, 2010)
2. Cao, X., et al.: Modeling Human Performance of Pen Stroke Gestures. In: CHI 2007, San Jose, Calif., USA, April 28–May 3 (2007)
3. Castellucci, et al.: Graffiti vs. Unistrokes: An Empirical Comparison. In: CHI 2008, Florence, Italy (2008)
4. Grossman, et al.: Hover Widgets: Using the Tracking State to Extend the Capabilities of Pen-Operated Devices. In: CHI 2006, Montreal, Quebec, Canada, April 22–27 (2006)
5. Hinckley, et al.: Design and Analysis of Delimiters for Selection-Action Pen Gesture Phrases in Scribolie. In: CHI 2005, Portland, Oregon, USA, April 2–7 (2005)
6. http://www.mobileorchard.com/
   iphone-circle-gesture-detection/

7. Liao, C., et al.: PapierCraft: A Gesture-Based Command System for Interactive Paper. In: TOCHI 2008, vol. 14(4) (January 2008)
8. Li, Y., et al.: Experimental Analysis of Mode Switching Techniques in Pen-based User Interfaces. In: SIGCHI Conf. on Human Factors in Computing Systems, April 02-07, pp. 461–470 (2005)
9. http://nuiman.com/log/view/lux/
10. Moyle, et al.: Analysing Mouse and Pen Flick Gestures. In: Proceedings of the SIGCHI-NZ Symposium On Computer-Human Interaction, New Zealand, July 11-12, pp. 19–24 (2002)
11. Opera web browser, http://www.opera.com (Cited March 27, 2010)
12. Roudaut, A., et al.: MicroRolls: Expanding Touch Screen Input Vocabulary by Distinguishing Rolls vs. Slides of the Thumb. In: CHI 2009, Boston, Mass, April 4–9 (2009)
13. Saund, E., et al.: Stylus Input and Editing Without Prior Selection of Mode. In: Symposium on User Interface Software and Technology, November 02–05 (2003)
14. Wobbrock, et al.: Maximizing the Guessability of Symbolic Input. In: CHI 2005, Portland, Oregon, USA, April 2-7 (2005)
15. Westerman, et al.: Writing using a Touch Sensor. US Patent Pub. No. US 2008/0042988 A1 (2007)
16. Zeleznik, R., et al.: Fluid Inking: Augmenting the Medium of Free-Form Inking with Gestures (2006)

A Recommendation Framework for Mobile Phones Based on Social Network Data

Alper Ozcan and Sule Gunduz Oguducu

Abstract. Nowadays, mobile phones are used not only for calling other people but also used for their various features such as camera, music player, Internet connection etc. Today's mobile service providers operate in a very competitive environment and they should provide other services to their customers than just call or SMS. One of such services may be to recommend items to their customers that match each customer's preferences and needs at the time the customer requests a recommendation. In this work, we designed a framework for an easy implementation of a recommendation system for mobile service providers. Using this framework, we implemented as a case study a recommendation model that recommends restaurants to the users based on the content filtering, collaborative filtering and social network of users. We evaluated the performance of our model on real data obtained from a Turkish mobile service company.

Keywords: Social network analysis, data mining, community detection, context-aware recommendation framework.

1 Introduction

In everyday life, we should make decisions from a variety of options. Decisions are sometimes hard to make because we do not have sufficient experience or knowledge. In this case, recommendations from our friends or groups that have experience would be very helpful. Software applications, named recommender systems, fulfill this social process. Recommender systems collect ratings from other users explicitly or implicitly and build a predictive model based on these ratings. While there have been significant research on recommender systems for Web users [5], there is little research for mobile phone users. Enabled by high technology with low costs, multi-talented mobile phones, that are not just phones anymore, are widely used. This makes mobile phones personal assistants for their users. An improved understanding of customer's habits, needs and interests can allow mobile service providers to profit in this competitive environment. Thus, recommender systems for mobile phones become an important tool in mobile environments. In this work, we designed a framework for an easy implementation

Alper Ozcan and Sule Gunduz Oguducu
Istanbul Technical University, Computer Engineering Department
34469, Maslak, Istanbul, Turkey
e-mail: ozcanalp@itu.edu.tr

Roger Lee (Ed.): SNPD 2010, SCI 295, pp. 139–149, 2010.
springerlink.com   © Springer-Verlag Berlin Heidelberg 2010

of a recommendation system for mobile phone users. The supported recommendation system is based on content filtering and collaborative filtering. The recommendation strategy is supported by social network approaches which enable to find communities or groups of people that know each other. The people in a social community are more connected to each other compared to the people in other communities of the network. Social theory tells us that social relationships are likely to connect similar people. If this similarity is used with a traditional recommendation model such as collaborative filtering, the resulting recommendation may be more suitable for the users' needs. With the increasing availability of user-generated social network data, it is now possible to drive recommendations more judiciously [11]. Thus, finding social network communities can provide great insights into recommendation systems. In this paper, since the recommendation application is designed to be used in mobile phone environment, the social network data are obtained from call data of mobile phone users. The motivation behind this work is that people that interact in their daily life through calling to each other may have similar preferences based on the assumption that they may spend some time together.

In this work, we implemented a framework of a recommendation system for mobile phone environment. The proposed framework consists of four steps: In the first step, the mobile phone user requests a recommendation. In the second step, the location of the user is determined based on the Global Positioning System (GPS). In the third step, recommendations are generated based on the content filtering, collaborative filtering and social network analysis methods. The last step is to send these recommendations to the user through short message services (SMS) or other interactive channels. Using this framework, we implemented a recommendation model that recommends restaurants to the users. We evaluated the performance of our model on real data obtained from a Turkish mobile service company.

The rest of the paper is organized as follows. In Section 2, the related works are summarized briefly. Section 3 introduces the techniques used in the recommendation framework. Section 4 and 5 present the framework architecture and recommendation methodology respectively. Finally, we describe the experimental results in Section 6 and conclude in Section 7.

2 Related Work

Various data mining techniques have been used to develop efficient and effective recommendation systems. One of these techniques is Collaborative Filtering (CF) [9]. A detailed discussion on recommender systems based on CF techniques can be found in [6]. A shortcoming of these approaches is that it becomes hard to maintain the prediction accuracy in a reasonable range due to the two fundamental problems: sparsity and cold-start problem [8]. Some hybrid approaches are proposed to handle these problems, which combines the content information of the items already rated. Content-based filtering techniques recommend items to the users based on the similarity of the items and the user's preferences.

Since their introduction, Internet recommendation systems, which offer items to the users on the basis of their preferences, have been exploited for many years.

However, there is little research on recommender systems for mobile phones. The recommendation systems in mobile phone environment differ in several aspects from traditional ones: The recommendation systems for mobile phones should interact with the environment where the user requested a recommendation. Context-aware recommendations are proposed to solve this problem [13]. The context is defined as any information that can be used to characterize the situation of any person or place that is considered relevant to the interaction between a user and an application. Also, a recent paper by Tung and Soo [13] implements a context-aware computing restaurant recommendation. They implement a prototype of an agent that utilizes the dialogue between user and the agent for modifying constraints given to the agent and recommends restaurants to the user based on the contexts in mobile environment. Another recommender system called critique-based recommender system is developed for mobile users [10]. This system is mainly based on the users' criticism of the recommended items to better specify their needs. However, when using mobile phones, usually with smaller screens and limited keypads, it is very difficult to get feedback from users about the recommended items. Moreover, the users are usually impatient when they are refining a recommendation list to select an item that fit their needs best at that time.

3 Background

In this section, we summarize the techniques used in this work to develop a recommendation framework for mobile phones. These techniques can be categorized as follows: content filtering, collaborative filtering, and community detection in a social network.

3.1 Content Filtering

Content-based recommendation systems recommend an item to a user based upon a description of the item and a profile of the user's interests. In content based recommendation every item is represented by a feature vector or an attribute profile. The features hold numeric or nominal values representing certain aspects of the item like color, price etc. Moreover, a content based system creates a user profile, and items are recommended for the user based on a comparison between item feature weights and those of the user profile [12]. In the case study evaluated in this paper, the pair-wise similarities between restaurants are calculated based on the feature vectors obtained from the attributes of these restaurants. User profile is also created for each user based on the features of the restaurants the user likes. The details of this method is explained in Section 5.

3.2 Collaborative Filtering

Collaborative recommendation systems identify users whose tastes are similar given a user and recommend to that user items which other similar users have liked. CF systems collect users' opinions on a set of objects, using ratings provided by the users or implicitly computed. In a user centric approaches rating matrix is constructed where each row represents a user and each column represents an item. CF systems predict a particular user's interest in an item using the rating matrix. It is often based on matching, in real-time, the current user's profile against similar records (nearest neighbors) obtained by the system over time from other users [1]. In this work, a rating matrix is constructed that each row represents a user and each column represents a restaurant. The similarities are computed between the active user and the other users using the rating matrix and the recommendations are then generated based on the preferences of the nearest neighbors for the active user. This method is also explained in Section 5 in detail.

3.3 Community Detection

In some applications the data can be represented as a graph structure $G = (V,E)$, for example: the web data, the call data etc. In that type of data, the set of nodes (V) represents the objects and the set of edges (E) represents the relation between these objects. *Undirected weighted graph* is a graph in which edges are not ordered pairs and a weight is assigned to each edge. Such weights might represent, for example, costs, lengths, etc. depending on the problem. A community in this graph structure is a group of objects that are highly connected to each other compared to the objects in other communities. There are several community detection algorithms for finding communities in undirected weighted graphs, such as greedy algorithm [2] and hierarchical clustering method [3]. Another community detection algorithm is a divisive approach proposed by Girvan and Newman [4]. It first finds the edges with the largest betweenness value (number of shortest paths passing through an edge) in decreasing order and then those edges are removed one by one in order to cluster graph into communities hierarchically. However, the complexity of this algorithm is $O(m^2n)$ where n and m are respectively the number of vertices and edges in the input graph.

In this work, the call data which consist of call records of users are represented as a graph. The identity and location information of users are also available in this data set. From this data, a call graph $G = (V,E)$ is obtained, where the set of nodes (V) represents the mobile phone users and the set of edges (E) represents the mobile calls. There is an undirected edge between to nodes w and u if w calls u or u calls w. The weight of this edge is the sum of the call durations between the nodes w and u. An algorithm based on random walks and spectral methods is employed for community detection. The reason of choosing this algorithm is that it works efficiently at various scales and community structures in a large network [7]. The complexity of this algorithm is reported as $O(n^2 log\ n)$ in most real-world complex networks. The resulting communities represent the set of people that communicate or call to each other frequently.

4 Framework Architecture

The overall framework architecture of the proposed recommendation system is shown in Fig. 1. The framework is designed as five main components: the

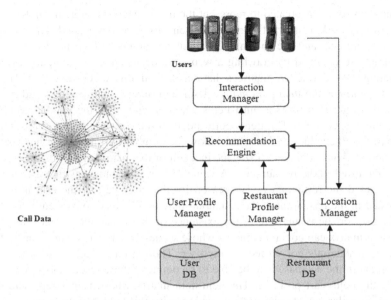

Fig. 1 The architecture of framework

interaction manager, the recommendation engine, the user profile manager, the restaurant profile manager and the location manager.

The interaction manager is the interface between users and recommendation engine. Its main role in the framework is to acquire users' requests and interact with users who request recommendation. It also delivers recommendation messages generated by recommendation engine to the user through SMS, MMS and mobile web pages. The location manager determines the current position of the user by using Global Positioning System (GPS) and notifies the location of the user to the recommendation engine. The location manager also retrieves location information about the nearby restaurants from the restaurant database which is formed from a data collection process that stores the features of restaurants. The user profile manager is capable of getting users' preferences based on the visit patterns of users. User database is the fundamental source for building the user profile. User profile describes user preferences for places. Restaurant profile manager retrieves the restaurants that satisfy the user's needs and preferences related to user's profile. The recommendation engine collects location and time information when it becomes a user request from the interaction manager. It is responsible to provide restaurant recommendations personalized to the user's needs and preferences at each particular request. Recommendations are produced based on the content filtering, collaborative filtering and social network analysis methods. Recommendations for an active user are generated based on the similarity between restaurants and the similarity between users in the same community in the social network.

Two data sources are used in this study. The first data source is the call data which are represented as a weighted undirected graph as mentioned in Section 3.3. The community detection process is executed by the User Profile Manager. The second data source is created to represent the users' interests on items. In the case study of this work, restaurants are used as items to be recommended. The features of the restaurants and the information about the number of visits of users to a restaurant are obtained by crawling a web site[1] designed for restaurants reviews and ratings. We define two matrices that derive from this data source (Fig. 2): the *user-item matrix*, the *item-item matrix*. User-item matrix, with m rows and n columns, is a matrix of users against items where m and n are the number of users and items respectively. The items correspond to the restaurants in the restaurant database and the values of that matrix show the number of times a user has visited a restaurant. Thus, a higher $r(i,j)$ value in a cell of this matrix shows that user u_i is more interested in the restaurant i_j. A value of 0 in $r(i,j)$ indicates that user u_i has never visited the restaurant i_j. The reason for that may be that the user u_i is not interested in the restaurant i_j; the restaurant may be not very conveniently located for the user or she is unaware of the existence of the restaurant. Besides this, the user-item matrix can contain columns whose values are all zeros which mean that there are restaurants not visited by any user. The reason to include such type of restaurants in the database is to be able to recommend these restaurants to users to handle the cold start problem. The item-item matrix whose values represent the pair-wise similarities of restaurants is used to solve this cold start problem.

To calculate the pair-wise similarities between restaurants in the Restaurant Profile Manager, a restaurant is first represented as a feature vector $Ri = (fi_1, fi_2, ..., fi_k)$, where each feature fi_j corresponds to an attribute of the restaurant Ri such as the type of the restaurant. For example, the vector Ri =({*nightclub*}, *30*,{*creditcard*},{*casual*},{*reservation not required*},*{Taksim}*) represents a restaurant, whose type is a *nightclub,* average cost is *30 euros*, the accepted method of payment is *credit card*, wearing apparel is *casual*, reservation prerequisite is *not required* and the restaurant is located in *Taksim*. The $w(k,j)$ value of the matrix, which shows the similarity between restaurant i_k and i_j, is calculated as follows:

$$w(k,j) = sim_{cos}(i_k, i_j) + |i_k(u) \cap i_j(u)| \tag{1}$$

where $sim_{cos}(i_k, i_j)$ is the cosine similarity between the feature vectors of restaurants i_k and i_j. $i_k(u)$ and $i_j(u)$ are the set of users that visit restaurant i_k and i_j respectively. Thus, $|i_k(u) \cap i_j(u)|$ is the number of users that visit both restaurant i_k and restaurant i_j.

$$sim(Ux, Uy) = \sum_{i=1}^{n} \sum_{j=1}^{n} r(x,i)r(y,j)w(i,j) \tag{2}$$

For each user u_x, a user profile $u_xprofile$ is constructed in the User Profile Manager as a feature vector $u_xprofile = (fu_{x1}, fu_{x2}, ..., fu_{xk})$ that obtained from all the feature vectors of the restaurants that the user has visited in the past. A feature fu_{xi} in the $u_xprofile$ corresponds to the ith feature in the feature vector of the restaurants and is computed as the mean of the ith feature of the restaurants that user u_x has

---

[1] http://www.mekanist.net

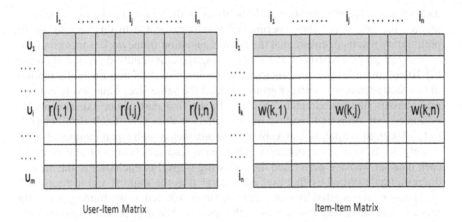

Fig. 2 The user-item matrix and the item-item matrix

visited in the past. In the case of a categorical feature, the mode of the feature, which is the most frequent value of that feature, is found. Given the user profile of an active user u_x and the similarity between a user u_y in the same community, a recommendation score $recScore(u_x, Ri)$ is computed by the Recommendation Engine for restaurant Ri that is visited by user u_y and not visited by u_x in the restaurant database as follows:

$$recScore(Ux, Ri) = \frac{1}{|S(Ri)|} \sum_{Uy \in S(Ri)} sim(Ux, Uy) sim(Uxprofile, Ri) \qquad (3)$$

In this equation, $sim(u_x profile, Ri)$ is the cosine similarity between u_x profile and the feature vector of restaurant Ri and $S(Ri)$ is the set of users that visit restaurant Ri and are in the same community as u_x. Each time a recommendation is requested from an active user, a similarity score is computed between the corresponding user and all other users in the same community based on their co-rated restaurants. In this method, the pair-wise similarities of users, restaurants and the recommendation scores of the restaurants are computed offline in order to decrease the online computing cost. When the user u_x requests a recommendation, the restaurant list of user u_x is filtered according to the location of this user and top-k restaurants that have the highest recommendation score $recScore(u_x, Ri)$ are recommended to that user.

5 Experimental Results

In the experimental part of this study, we used two data sources: Call graph obtained from a Turkish Mobile Service Provider and a restaurant data set. The restaurant database consists of 725 restaurants located in Istanbul. The call graph is constructed for 551 users of a Turkish Mobile Service Provider. Approximately 70% of the restaurants are randomly selected as the training set and the remaining part as the test set. The restaurants in the training set are used for calculating the values of user-item and item-item matrices. Please note that the users in the call data set are randomly matched with the users in the restaurant data set due to the privacy reasons. Thus, the experimental results presented in this section are the preliminary results to ensure that the recommendation framework works properly.

The call graph consists of 551 users. Using the community detection algorithm [7] 10 communities are constructed. The community detection method employed in this study tends to construct groups around the same size. Thus, every community consists of approximately 55 users. In the first part of our experiments, we evaluated our proposed method on 10 different data sets. Each data set (*DSi*, *i*=1…,10, in the experimental Table 1.) is constructed with equal number of users, namely 30, from every community. Thus, each data set consists of 300 users. On each data set we have performed the following steps: For each user in the data set, the restaurants visited by that user are separated as the training and test sets. 70% of the restaurants visited by a user are randomly selected as the training set of this user and the remaining restaurants as the test set. The user profile ($u_x profile$) of a user u_x is constructed based on the restaurants in the training set of the corresponding user. Naming the user to make recommendation for as active user, 5 most similar users are selected with highest similarity score (Eq. 2) in the same community with the active user. The restaurants that have been visited by those 5 users (but not by active user) with highest recommendation scores (Eq. 3) are recommended to the active user. After all recommendations are done for every user, the Precision, Recall and F-Measure values of each data set are calculated by the formulas given below.

$$Precision = \frac{|Recommended\ restaurants \cap Restaurants\ in\ Test\ Set\ |}{|Recommended\ restaurants|} \qquad (4)$$

$$Recall = \frac{|Recommended\ restaurants \cap Restaurants\ in\ Test\ Set\ |}{|Restaurants\ in\ Test\ Set|} \qquad (5)$$

$$F-measure = \frac{2 \times Precision \times Recall}{Precision + Recall} \qquad (6)$$

If the recommended restaurant is in the test set of the user u_x, then the recommendation is marked as correct for user u_x. To calculate precision value for each dataset, the total number of correct recommendations for every user in that dataset is divided to the total number of recommendations made by the system (Eq. 4). On the other hand, to calculate recall value of each dataset, the total number of correct recommendations for every user in that dataset is divided to the total number of restaurants in the test set of these users (Eq. 5).

In the second part of the experiments, we compared our model with a Random User Model (RUM). The reason to compare the results of the framework with a random model is because of the matching strategy of users in the call data set and the restaurant data set. As mentioned before, a user in the call data is matched randomly with someone in the restaurant data set. We evaluated RUM method on 10 different data sets. Each data set (*DSRi*, *i*=1…,10, in the experimental Table 2.) is constructed with equal number of users, namely 30, from the call graph. For this set of experiments, data sets are created by selecting randomly 30 users from the call graph. On each data set we performed the following experiment: For each user, 5 random users are selected to make recommendations. For the active user, the restaurants that have been visited by 5 random users (again not visited by active user) are recommended. After all recommendations are done for every user, the Precision, Recall and F-Measure values are calculated. This experiment was run 10 times for each data set. The average values of 10 runs become Precision, Recall and F-Measure values of the data set. Table 1 and Table 2 present these results.

As it can be seen from the Table 1 and Table 2, our proposed recommendation model has a significantly better result than RUM. These experiments justify the use of social network approaches, which enable to find social communities, increases the accuracy of the recommendation model. The average precision of recommendation messages, which are delivered according to users' context and social relationships, was over 0.63%.

The advantages of the proposed method are the following: (1) Using the social network, we narrowed the recommendation search space which resulted in simplification of the calculations. (2) As the results also indicate, our prediction model produces better results than RUM in terms of Precision, Recall and F-measure. (3) The online computing cost is low. (4) Using social network structure the interaction between users and framework is very limited.

Table 1 Average results of proposed method.

Proposed Method	Precision	Recall	F-Measure
DS1	0.6488	0.2092	0.3164
DS2	0.5858	0.1926	0.2901
DS3	0.6697	0.2188	0.3299
DS4	0.6796	0.2250	0.3381
DS5	0.6059	0.2013	0.3022
DS6	0.6082	0.1996	0.3006
DS7	0.6165	0.2041	0.3067
DS8	0.6199	0.2038	0.3068
DS9	0.6666	0.2201	0.3309
DS10	0.6682	0.2192	0.3302
Average	0.6369	0.2094	0.3152

Table 2 Average results of random user model.

Random User Model	Precision	Recall	F-Measure
DSR1	0.0994	0.0324	0.0489
DSR2	0.1320	0.0420	0.0637
DSR3	0.1301	0.0416	0.0631
DSR4	0.1864	0.0598	0.0906
DSR5	0.0668	0.0212	0.0322
DSR6	0.1883	0.0602	0.0913
DSR7	0.0695	0.0221	0.0336
DSR8	0.1753	0.0571	0.0861
DSR9	0.2490	0.0859	0.1277
DSR10	0.1424	0.0452	0.0686
Average	0.1439	0.0468	0.0706

6 Conclusion

In this paper, we presented a framework for a prototype of a mobile recommendation system. We also demonstrated how the recommendation system can incorporate context awareness, user preference, collaborative filtering, and social network approaches to provide recommendations. Location and time are defined as contextual information for recommendation framework. The recommendation strategy is supported by social network approach which enables to find social relationships and social communities. In this paper, the main idea is to develop a recommendation system based on the call graph of mobile phone users. The social network communities are used in order to provide insight into recommendation systems. Using social relationships with traditional recommendation models such as collaborative filtering provides to find suitable choices for the users' needs. The intuition behind this idea is that groups of people who call each other frequently may spend some time together and may have similar preferences. In order to evaluate the performance of the framework, a call graph from a Mobile Service Provider and restaurant dataset are used in the case study of this work. Our preliminary experimental results of our proposed method are promising. Recommendation techniques based on social network information and CF can enhance and complement each other so that they help users to find relevant restaurants according to their preferences.

Currently, we are working on adapting the recommendation framework to recommend different types of items. Since the design of the recommendation framework is flexible and extendible, it can be easily tailored to the specific application needs. It can be also used to recommend items such as hotels, cinemas, hospitals, movie and music selections etc.

References

1. Breese, J.S., Heckerman, D., Kadie, C.: Empirical Analysis of Predictive Algorithms for Collaborative Filtering. Technical Report MSR-TR-98-12, Microsoft Research, Redmond (1998)
2. Caflisch, A., Schuetz, P.: Efficient modularity optimization by multistep greedy algorithm and vertex mover refinement. Physical Review E, Zurich (2008)
3. Donetti, L., Munoz, M.: Detecting Network Communities: a new systematic and efficient algorithm. J. Stat. Mech. Theor. Exp. (2004)
4. Girvan, M., Newman, M.E.J.: Community structure in social and biological networks. Proceedings of the National Academy of Sciences 99(12) (2002)
5. Golbeck, J.: Generating Predictive Movie Recommendations from Trust in Social Networks. In: Stølen, K., Winsborough, W.H., Martinelli, F., Massacci, F. (eds.) iTrust 2006. LNCS, vol. 3986, pp. 93–104. Springer, Heidelberg (2006)
6. Herlocker, J.L., et al.: Evaluating collaborative filtering recommender systems. ACM Transactions on Information Systems 22(1), 5–53 (2004)
7. Latapy, M., Pons, P.: Computing communities in large networks using random walks. In: Yolum, p., Güngör, T., Gürgen, F., Özturan, C. (eds.) ISCIS 2005. LNCS, vol. 3733, pp. 284–293. Springer, Heidelberg (2005)

8. Melville, P., Mooney, R.J., Nagarajan, R.: Content-Boosted Collaborative Filtering for Improved Recommendations. In: Proceedings of the Eighteenth National Conference on Artificial Intelligence, Edmonton, Canada (2002)
9. Pazzani, M.S.: A Framework for Collaborative, Content-Based and Demographic Filtering. Artificial Intelligence Review 13, 393–408 (1999)
10. Ricci, F., Nguyen, Q.N.: Critique-Based Mobile Recommender Systems. ÖGAI Journal 24(4) (2005)
11. Seth, A., Zhang, J.: A Social Network Based Approach to Personalized Recommendation of Participatory Media Content. In: International Conference on Weblogs and Social Media, Seattle, Washington (2008)
12. Sobecki, J.: Implementations of Web-based Recommender Systems Using Hybrid Methods. International Journal of Computer Science & Applications 3, 52–64 (2006)
13. Tung, H., Soo, V.: A Personalized Restaurant Recommender Agent for Mobile E- Service. In: Proceedings of IEEE International Conference on e-Technology, e-Commerce and e-Service (2004)

A Reputation-based Framework for Mobile Phones that Should Prevent Data... 159

Language Modeling for Medical Article Indexing

Jihen Majdoubi, Mohamed Tmar, and Faiez Gargouri

Abstract. In the medical field, scientific articles represent a very important source of knowledge for researchers of this domain. But due to the large volume of scientific articles published on the web, an efficient detection and use of this knowledge is quite a difficult task.

In this paper, we propose a contribution for conceptual indexing of medical articles by using the MeSH (Medical Subject Headings) thesaurus, then we propose a tool for indexing medical articles called SIMA (System of Indexing Medical Articles) which uses a language model to extract the MeSH descriptors representing the document.

Keywords: Medical article, Semantic indexing, Language models, MeSH thesaurus.

1 Introduction

In the medical field, scientific articles represent a very important source of knowledge for researchers of this domain. The researcher usually needs to deal with a large amount of scientific and technical articles for checking, validating and enriching of his research work.

The needs expressed by these researchers can be summarized as follows:

- The medical article may be a support of validation of experimental results: the researcher needs to compare his clinical data to already existing data sets and to reference knowledge bases to confirm or invalidate his work.
- The researcher may want to obtain information about a particular disease (manifestations, symptoms, precautions).

Jihen Majdoubi, Mohamed Tmar, and Faiez Gargouri
Multimedia InfoRmation system and Advanced Computing Laboratory,
Higher Institute of Computer Science and Multimedia, Sfax-Tunisia
e-mail: `majdoubi_jihene@yahoo.fr`, `mohamed.tmar@isimsf.rnu.tn`,
`Faiez.Gargouri@fsegs.rnu.tn`

Roger Lee (Ed.): SNPD 2010, SCI 295, pp. 151–161, 2010.

This kind of information is often present in electronic biomedical resources available through the Internet like CISMEF[1] and PUBMED[2]. However, the effort that the user puts into the search is often forgotten and lost.

Thus much more "intelligence" should be embedded to Information Retrieval System (IRS) in order to be able to understand the meaning of the word: it can be carried out by associating to each document an index based on a semantic resource describing the domain.

In this paper, we are interested in this field search. More precisely, we propose our contribution for conceptual indexing of medical articles by using the language modeling approach.

The remainder of this article is organized as follows. In section 2, we started out with over viewing related work according to indexing medical articles. Following that, we detail our conceptual indexing approach in Section 3. An experimental evaluation and comparison results are discussed in section 4. Finally section 5 presents some conclusions and future work directions.

2 Previous Work

Automatic indexing of the medical literature has been investigated by several researchers. In this section, we are only interested in the indexing approach using the MeSH thesaurus.

[1] proposes a tool called MAIF (MesH Automatic Indexer for French) which is developed within the CISMeF team. To index a medical ressource, MAIF follows three steps: analysis of the resource to be indexed, translation of the emerging concepts into the appropriate controlled vocabulary (MeSH thesaurus) and revision of the resulting index.

In [2], the authors proposed the MTI (MeSH Terminology Indexer) used by NLM to index English resources. MTI results from the combination of two MeSH Indexing methods: MetaMap Indexing (MMI) and a statistical, knowledge-based approach called PubMed Related Citations (PRC).

The MMI method [3] consists of discovering the Unified Medical Language System (UMLS) concepts from the text. These UMLS concepts are then refined into MeSH terms.

The PRC method [4] computes a ranked list of MeSH terms for a given title and abstract by finding the MEDLINE citations most closely related to the text based on the words shared by both representations.

Then, MTI combines the results of both methods by performing a specific post processing task, to obtain a first list. This list is then devoted to a set of rules designed to filter out irrelevant concepts. To do so, MTI provides three levels of filtering depending on precision and recall: the strict filtering, the medium filtering and the base filtering.

---

[1] http://www.chu-rouen.fr/cismef/
[2] http://www.ncbi.nlm.nih.gov/pubmed/

Nomindex [5] recognizes concepts in a sentence and uses them to create a database allowing to retrieve documents. Nomindex uses a lexicon derived from the ADM (Assisted Medical Diagnosis) [6] which contains 130.000 terms.

First, document words are mapped to ADM terms and reduced to reference words. Then, ADM terms are mapped to the equivalent French MeSH terms, and also to their UMLS Concept Unique Identifier. Each reference word of the document is then associated with its corresponding UMLS. Finally a relevance score is computed for each concept extracted from the document.

[7] showed that the indexing tools cited above by using the controlled vocabulary MeSH, increase retrieval performance.

These approachs are based on the vector space model. We propose in the next section our approach for the medical article indexing which is based on the language modeling.

3   Our Approach

The basic approach for using language models for IR assumes that the user has a reasonable ideal of the terms that are likely to appear in the ideal document that can satisfy his/her information need, and the query terms the user chooses can distinguish the ideal document from the rest of the collection [8]. The query is generated as the piece of text representative of the ideal document. The task of the system is then to estimate, for each of the documents in the collection, which is most likely to be the ideal document.

Our work aims to determine for each document, the most representative MeSH descriptors. For this reason, we have adapted the language model by substituting the query by the Mesh descriptor. Thus, we infer a language model for each document and rank Mesh descriptor according to our probability of producing each one given that model. We would like to estimate $P(Des|M_d)$, the probability of the Mesh descriptor given the language model of document d.

Our indexing methodology as shown by figure 1, consists of three main steps: (a) Pretreatment, (b) concept extraction and (c) generation of the semantic core of document.

We present the architecture components in the following subsections.

3.1   MeSH Thesaurus

The structure of MeSH is centered on descriptors, concepts, and terms.

- Each term can be either a simple or a composed term.
- A concept is viewed as a class of synonymous terms, one of then (called Preferred term) gives its name to the concept.
- A descriptor class consists of one or more concepts where each one is closely related to each other in meaning.

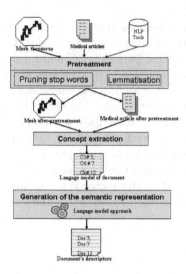

Fig. 1 Architecture of our proposed approach

Kyste du cholédoque [Descriptor]
Kyste du cholédoque [Concept, Preferred]
Kyste du cholédoque [Term, Preferred]
Kyste du canal cholédoque [Term]
Kyste du cholédoque de type V [Concept, Narrower]
Kyste du cholédoque de type V [Term, Preferred]
Kyste du cholédoque de type 5 [Term]
Kyste du cholédoque intrahépatique [Term]

Fig. 2 Extrait of MeSH

Each descriptor has a preferred concept. The descriptor's name is the name of the preferred concept. Each of the subordinate concepts is related to the preferred concept by a relationship (broader, narrower).

As shown by figure 2, the descriptor "*Kyste du cholédoque*" consists of two concepts and five terms. The descriptor's name is the name of its preferred concept. Each concept has a preferred term, which is also said to be the name of the Concept. For example, the concept "*Kyste du cholédoque*" has two terms "*Kyste du cholédoque*" (preferred term) and "*Kyste du canal cholédoque*".

As in the example above, the concept "*Kyste du choldoque de type V*" is narrower to than the preferred concept "*Kyste du canal cholédoque*".

3.2 Pretreatment

The first step is to split text into a set of sentences. We use the Tokeniser module of GATE [9] in order to split the document into tokens, such as numbers, punctuation, character and words. Then, the TreeTagger [10] stems these tokens to assign a grammatical category (noun, verb...) and lemma to

> T002648 enfant
> T000368 sujet âgé
> T035922 anticorps hépatite
> T014780 étude clinique

Fig. 3 Example of terms

each token. Finally, our system prunes the stop words for each medical article of the corpus. This process is also carried out on the MeSH thesaurus.

Thus, the output of this stage consists of two sets. The first set is the article's lemma, and the second one is the list of lemma existing in the MeSH thesaurus.

3.3 Concept Extraction

This step consists of extracting single word and multiword terms from texts that correspond to MeSH concepts. So, SIMA processes the medical article sentence by sentence. Indeed, in the pretreatment step, each lemmatized sentence S is represented by a list of lemma ordered in S as they appear in the medical article. Also, each MeSH term t_j is processed with TreeTagger in order to return its canonical form or lemma. Let: $S = (l_1, l_2, \ldots, l_n)$ and $(t_j = (att_{j1}, att_{j2}, \ldots, att_{jk}))$. The terms of a sentence S_i are:

$$Terms(S_i) = \{T, \forall att \in T, \exists l_{ij} \in S_i, att = l_{ij}\}$$

For example, let us consider the lemmatized sentence S_1 given by:

$$S_1 = (\text{étude}, enfant, agé, sujet, anticorps, virus, hépatite\}.$$

If we consider the set of terms shown by figure 3, this sentence contains three different terms: (i) *enfant*, (ii) *sujet agé* and (iii) *anticorps hépatite*. The term *étude clinique* is not identified because the word *clinique* is not present in the sentence S_1.

Thus:

$$Terms(S_1) = \{enfant, sujet\ agé, anticorps\ hépatite\}.$$

A concept c_i is proposed to the system like a concept associated to the sentence S $(Concepts(S))$, if at least one of its terms belongs to $Terms(S)$.

For a document d composed of n sentences, we define its concepts (Concepts(d)) as follows:

$$Concepts(d) = \bigcup_{i=1}^{n} Concepts(S_i) \tag{1}$$

Given a concept c_i of $Concepts(d)$, its frequency in a document d ($f(c_i, d)$) is equal to the number of sentences where the concept is designated as $Concepts(S)$. Formally:

$$f(c_i, d) = \left| \sum_{c_i \in Concepts(S_j)} S_j \in d \right| \qquad (2)$$

3.4   Generation of the Semantic Core of Document

To determine the MeSH descriptors from documents, we estimated a language model for each document in the collection and for a MeSH descriptor we rank the documents with respect to the likelihood that the document language model generates the MeSH descriptor. This can be viewed as estimating the probability $P(d|des)$.

To do so, we used the language model approach proposed by [11].

For a collection D, document d and MeSH descriptor (des) composed of n concepts:

$$P(d|des) = P(d). \prod_{c_j \in des} (1 - \lambda).P(c_j|D) + \lambda.P(c_j|d) \qquad (3)$$

We need to estimate three probabilities:

1. $P(d)$: the prior probability of the document d:

$$P(d) = \frac{|concepts(d)|}{\sum_{d' \in D} |concepts(d')|} \qquad (4)$$

2. $P(c|D)$: the probability of observing the concept c in the collection D:

$$P(c|D) = \frac{f(c, D)}{\sum_{c' \in D} f(c', D)} \qquad (5)$$

where $f(c, D)$ is the frequency of the concept c in the collection D.

3. $P(c|d)$: the probability of observing a concept c in a document d:

$$P(c|d) = \frac{cf(c, d)}{|concepts(d)|} \qquad (6)$$

Several methods for concept frequency computation have been proposed in the literature. In our approach, we applied the weighting concepts method (CF: Concept Frequency) proposed by [12].

So, for a concept c composed of n words, its frequency in a document depends on the frequency of the concept itself, and the frequency of each sub-concept. Formally:

$$cf(c,d) = f(c,d) + \sum_{sc \in subconcepts(c)} \frac{length(sc)}{length(c)} \cdot f(sc,d) \qquad (7)$$

with:

- $Length(c)$ represents the number of words in the concept c.
- $subconcepts(c)$ is the set of all possible concepts MeSH which can be derived from c.

For example, if we consider a concept "bacillus anthracis", knowing that "bacillus" is itself also a MeSH concept, its frequency is computed as:

$$cf(bacillus\ anthracis) = f(bacillus\ anthracis) + \frac{1}{2} \cdot f(bacillus)$$

consequently:

$$P(d|des) = \frac{|concepts(d)|}{\sum\limits_{concepts(d') \in D} |concepts(d')|}$$

$$\cdot \prod_{c \in des} (1 - \lambda) \cdot \frac{f(c, D)}{\sum\limits_{c' \in D} f(c', D)}$$

$$+ \lambda \cdot \left(\frac{f(c,d) + \sum\limits_{sc \in subconcepts(c)} \frac{length(sc)}{length(c)} \cdot f(sc,d)}{|concepts(d)|} \right) \qquad (8)$$

4 Empirical Results

4.1 Test Corpus

To evaluate our indexing approach we are based on the same corpus used by [13]. This corpus is composed of 82 resources randomly selected in the CIS-MeF catalogue. It contains about 235,000 words altogether, which represents about 1.7 Mb.

Each document of this corpus has been manually indexed by five professional indexers in the CISMeF team in order to provide a description called "notice". Figure 4 shows an example of notice for resource "diabte de type 2".

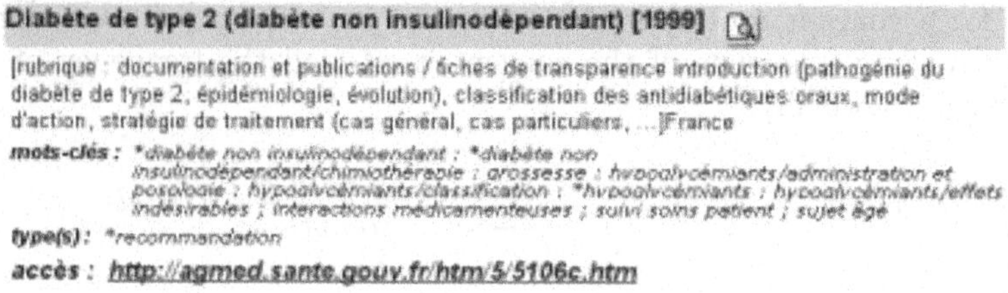

Diabète de type 2 (diabète non insulinodépendant) [1999]

[rubrique : documentation et publications / fiches de transparence introduction (pathogénie du diabète de type 2, épidémiologie, évolution), classification des antidiabétiques oraux, mode d'action, stratégie de traitement (cas général, cas particuliers, ...]France

mots-clés : **diabète non insulinodépendant ; *diabète non insulinodépendant/chimiothérapie ; grossesse ; hypoglycémiants/administration et posologie ; hypoglycémiants/classification ; *hypoglycémiants ; hypoglycémiants/effets indésirables ; interactions médicamenteuses ; suivi soins patient ; sujet âgé*

type(s) : **recommandation*

accès : http://agmed.sante.gouv.fr/htm/5/5106c.htm

Fig. 4 CISMeF notice for resource *"Diabte de type 2"*

A notice is mainly composed of:

- General description: title, authors, abstract,....
- Classification: MeSH descrptors that describe article content.

In this evaluation, the notice (manual indexing) is used as a reference.

4.2 Experimental Process

Our experimental process is undretaken as follows:

- Our process begins by dividing each article into a set of sentences. Then, a lemmatisation of the corpus and the Mesh terms is ensured by TreeTagger[14]. After that, a filtering step is performed to remove the stop words.
- For each sentence S_i, of a test corpus, we determine the set $concepts(S_i)$.
- For a document d and for each MeSH descriptor des_i, we calculate $P(d|des_i)$.
- For each document d, the MeSH descriptors are rankeded by decreasing scores $P(d|des_i)$.

4.3 Evaluation and Comparison

To evaluate our indexing tool, we carry out an experiment which compares, on the test corpus set with two MeSH indexing systems: MAIF (MeSH Automatic Indexing for French) and NOMINDEX developed in the CISMeF team. For this evaluation, we used three measures: precision, recall and F-measure.

Precision corresponds to the number of indexing descriptors correctly retrieved over the total number of retrieved descriptors.

Recall corresponds to the number of indexing descriptors correctly retrieved over the total number of expected descriptors.

F-measure combines both precision and recall into a single measure.

$$Recall = \frac{TP}{TP + FN} \tag{9}$$

$$Precision = \frac{TP}{TP + FP} \tag{10}$$

Where:

- TP: (true positive) is the number of MeSH descriptors correctly identified by the system and found in the manual indexing.
- FN: (false negative) is the MeSH descriptors that the system failed to retrieve in the corpus.
- FP: (false positive) is the number of MeSH descriptors retrieved by the system but not found in the manual indexing.

$$F - measure = \frac{1}{\alpha \times \frac{1}{Precision} + (1 - \alpha) \times \frac{1}{Recall}} \tag{11}$$

where $\alpha = 0, 5$.

Table 1 shows the precision and recall obtained by NOMINDEX, MAIF and SIMA at fixed ranks 1,4, 10 and 50 on the test Corpus.

Table 1 Precision and recall of NOMINDEX, MAIF and SIMA

Rank	NOMINDEX (P/R)	MAIF (P/R)	SIMA (P/R)
1	13,25/2,37	45,78/7,42	32,28/6,76
4	12,65/9,20	30,72/22,05	27,16/21,56
10	12,53/22,55	21,23/37,26	22,03/40,62
50	6,20/51,44	7,04/48,50	11,19/62,08

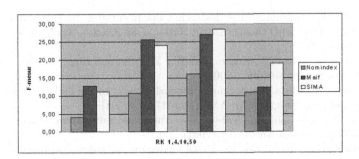

Fig. 5 Comparative results of F-measure

Figure 5 gives the F-measure value generated by NOMINDEX, MAIF and SIMA at fixed ranks 1,4, 10 and 50.

By examining the figure 5, we can notice that the least effective results come from NOMINDEX with a F-measure value equal to $4, 02$ in rank 1, $10, 65$ in rank 4, $16, 11$ in rank 10 and $11, 07$ in rank 50.

MAIF generates the best performance results with a F-measure value equal to $12, 77$ % at rank 1 and $24, 03$ at.

At ranks 10 and 50, the best results was achieved by our system SIMA with respectively F-measure values $28, 56$ and $18, 96$ at rank 4.

5 Conclusion

The work developed in this paper outlined a concept language model using the Mesh thesaurus for representing the semantic content of medical articles.

Our proposed conceptual indexing approach consists of three main steps.

At the first step (Pretreatment), being given an article, MeSH thesaurus and the NLP tools, the system SIMA extracts two sets: the first is the article's lemma, and the second is the list of lemma existing in the the MeSH thesaurus.

At step 2, these sets are used in order to extract the Mesh concepts existing in the document. After that, our system interpret the relevance of a document d to a MeSH descriptor des by measuring the probability of this descriptor to be generated by a document language ($P(d|des_i)$).

Finally, the MeSH descriptors are rankeded by decreasing score $P(d|des_i)$.

An experimental evaluation and comparison of SIMA with MAIF and NO-MINDEX confirm the well interest to use the language modeling approach in the conceptual indexing process. However, many future work directions can be considered. We must think to integrate a kind of semantic smoothing into the langage modeling approach.

References

1. Neveol, A.: Automatisation des taches documentaires dans un catalogue de santté en ligne. PhD thesis, Institut National des Sciences Appliquées de Rouen (2005)
2. Aronson, A., Mork, J., Rogers, C.S.W.: The nlm indexing initiative's medical text indexer. In: Medinfo. (2004)
3. Aronson, A.: Effective mapping of biomedical text to the umls metathesaurus: The metamap program. In: AMIA, pp. 17–21 (2001)
4. Kim, W.: Aronson, A., Wilbur, W.: Automatic mesh term assignment and quality assessment. In: AMIA (2001)
5. Pouliquen, B.: Indexation de textes médicaux par indexation de concepts, et ses utilisations. PhD thesis, Université Rennes 1 (2002)
6. Lenoir, P., Michel, R., Frangeul, C., Chales, G.: Réalisation, développement et maintenance de la base de données a.d.m. In: Médecine informatique (1981)
7. Neveol, A., Pereira, S., Kerdelhué, G., Dahamna, B., Joubert, M., Darmoni, S.: Evaluation of a simple method for the automatic assignment of mesh descriptors to health resources in a french online catalogue. In: MedInfo. (2007)
8. Ponte, M., Croft, W.: A language modeling approach to information retrieval. In: ACM-SIGIR Conference on Research and Development in Information Retrieval, pp. 275–281 (1998)

9. Cunningham, M., Maynard, D., Bontcheva, K., Tablan, V.: Gate: A framework and graphical development environment for robust nlp tools and applications. ACL (2002)
10. Schmid, H.: Probabilistic part-of-speech tagging using decision trees. In: International Conference on New Methods in Language Processing, Manchester (1994)
11. Hiemstra, D.: Using Language Models for Information Retrieval. PhD thesis, University of Twente (2001)
12. Baziz, M.: Indexation conceptuelle guidée par ontologie pour la recherche d'information. PhD thesis, Univ. of Paul sabatier (2006)
13. Névéol, A., Mary, V., Gaudinat, A., Boyer, C., Rogozan, A., Darmoni, S.J.: A benchmark evaluation of the french mesh indexers. In: Miksch, S., Hunter, J., Keravnou, E.T. (eds.) AIME 2005. LNCS (LNAI), vol. 3581, pp. 251–255. Springer, Heidelberg (2005)
14. Schmid, H.: Probabilistic part-of-speech tagging using decision trees. In: International Conference on New Methods in Language Processing, Manchester (1994)

Author Index